Innovating Out *of* Crisis

Innovating Out of Crisis

How Fujifilm Survived (and Thrived) As Its Core Business Was Vanishing

SHIGETAKA KOMORI
CHAIRMAN AND CEO, FUJIFILM HOLDINGS CORPORATION

Stone Bridge Press • *Berkeley, California*

Published by
Stone Bridge Press
P. O. Box 8208, Berkeley, CA 94707
tel 510-524-8732
sbp@stonebridge.com • www.stonebridge.com

Japanese edition published in 2013 as *Tamashii no Keiei* by Toyo Keizai Inc., Tokyo, Japan.

© 2015 Shigetaka Komori.

All rights reserved.

No part of this book may be reproduced in any form without permission from the publisher.

Printed in Japan.

10 9 8 7 6 5 4 3 2 1   2019 2018 2017 2016 2015

LIBRARY OF CONGRESS CATALOGING-IN-PUBLICATION DATA
Komori, Shigetaka.
  [Tamashii no keiei. English]
  Innovating out of crisis : how Fujifilm survived (and thrived) as its core business was vanishing / Shigetaka Komori.
     pages cm
  "Japanese edition published in 2013 as Tamashii no Keiei by Toyo Keizai Inc., Tokyo, Japan."
  ISBN 978-1-61172-023-5 (p-book)
  ISBN 978-1-61172-915-3 (e-book)
  1. Komori, Shigetaka. 2. Fuji Shashin Firumu Kabushiki Kaisha. 3. Photographic film industry—Japan—Management. 4. Photographic industry—Japan—Management. 5. Strategic planning—Japan—Case studies. I. Title.
  HD9708.5.F544F85513 2015
  338.7'6177153240952—dc23

                                  2014037584

CONTENTS

Introduction   11

PART ONE: FIGHTING FOR FUJIFILM

1   *The Core Business Vanishes*   17

   The Coming Crisis   19
   Technology-Oriented Fujifilm   20
   The Kodak Giant   22
   First Attempts at Diversification   23
   Overseas Development and Foreign Encroachment   27
   The Approaching Digital Era   29
   Three Strategies for the Digital Age   33
   Brave Enough to Recognize Reality?   38
   The Crisis Arrives   40

   Sidebar: Digital Minilabs: A Godsend for Photo Stores   42

2 A Second Foundation 47

My Return from Europe 49

Not Just to Survive, But to Thrive as a First-Rate Enterprise 52

Reorganizing and Consolidating 55

Pushing Relentless Reform, But with Consideration for All 57

Preserving the Culture of Photography 59

FUJITAC and the Market Growth of Liquid Crystal Television 61

Needs and Technology: Searching for New Business 65

Not Just Success, But Long-term Success 67

Healthcare's Importance in the Twenty-first Century 72

The Rationale behind Cosmetics 73

A Full-Scale Entry into Pharmaceuticals 76

Mergers and Acquisitions Provide a Head Start 79

Creating a New Center for Interdisciplinary Research 82

Ongoing R&D Investment of ¥200 Billion a Year 86

Transition to a Holding Company 88

A New Name: "Fujifilm" 90

Our Best Performance in History—Then the Global Collapse of 2008 *91*

A Second Companywide Restructuring *93*

One Misfortune Follows Another: The Strong Yen *96*

Back on the Path to Growth *98*

What's Next?—After the Trunk and Limbs Come the Branches and Leaves *101*

Creating a Company That Can Create Change *103*

The Difference between Kodak and Fujifilm *105*

> *Sidebar: Disaster Reconfirms the Cultural Value of Photography 109*

PART TWO: MANAGING FOR VICTORY

3   *Managing in Times of Crisis* *115*

Consensus Leaders Are Useless *117*

Four Steps for Managers in Times of Crisis *119*

Understanding the Present Situation with Limited Information *121*

Reading the Flow of Events and Predicting the Future *124*

Applying Universal Laws Outside Your Area of Expertise *127*

Three Ways of Misreading the Present and Future *128*

Deciding Priorities and Drawing Up Realistic Plans *131*

Dynamism and Speed *132*

The Need for Muscle Intelligence *134*

Even When You Hesitate, Make It a Success *135*

Keeping Refreshed and Invigorated *137*

Without Communication from the Top, the Organization Won't Budge *138*

Leading Is More Important Than Thinking about How to Lead *141*

> *Sidebar: Number Two Uses a Bamboo Sword, Number One Uses Steel 143*

4   A Battle That Cannot Be Lost *145*

All Life Is a Battle to Be Won or Lost *147*

Postwar Japan Teaches Me the Wretchedness of Losing *149*

Building a Bedrock of Strength to Escape Defeat *151*

The Whole Body Theory of Business *153*

Without Gentleness and a Cause, "Winning" and "Strength" Are Meaningless *158*

Winning through Understanding International Behavior and Manners   *159*

Winning Intelligently, Honestly, and with Spirit   *162*

> *Sidebar: Books I Have Read to Build a Foundation of Strength   165*

5   *Those Who Put the Company First Are Those Who Truly Grow*   *171*

The Company Is Not a Classroom   *173*

Learn from Whatever Comes Your Way   *175*

Work with a Sense of Ownership   *177*

Take Whatever You Do Seriously and See It Through   *180*

Before Relying on Others, Ask Yourself What You Have Done   *182*

Without Changing Reality, There Is No Progress   *184*

Why Some Senior Managers Don't Grow   *185*

> *Sidebar: Not Plan-Do-Check-Action, But See-Think-Plan-Do   187*

6   *The Way Forward in a Global Age*   *191*

Japan's Manufacturing Sector Is Not Losing Ground   *193*

The Slow Economy Due to a Strong Yen   *195*

Separating TPP and Agricultural Issues   *200*

Issues for Japan: The High Cost of Corporate SG&A   *202*

Issues for Japan: Deterioration in the Ability to Execute   *205*

Issues for Japan: Blurring Responsibility   *207*

Japanese Technology: Still a Source of Pride   *208*

Teaching Children the Importance of Competition   *210*

From Backward, Inward, and Downward to Forward, Outward, and Upward   *214*

Conclusion   *217*

INTRODUCTION

In the year 2000, photosensitive materials such as color photographic film accounted for sixty percent of Fujifilm's sales and two-thirds of its profits. The disappearance of that market in roughly the blink of an eye, as digital photography grew dominant, was for us an earth-shattering event. Photosensitive materials, our core moneymaker, had fallen into the red in just four or five short years.

It was during this, the deepest crisis in our company's history, that I was appointed president.

The question facing me now was how to survive this unprecedented situation and save the company. If I failed, it would mean disaster for Fujifilm; and for me personally, at this final stage of my career as a company man, it would mean my whole life had been a failure.

* * *

Photography is an indispensable part of human culture, and Fujifilm takes it as its mission to protect the photographic culture. We cannot stop producing film simply because the market has shrunk.

But if the demand for photographic film drastically dwindled, our production facilities and sales organizations around the world would become a huge fixed cost weighing us down. If Fujifilm could not sustain this burden and went under, the lives of some seventy thousand employees and their families would be shattered.

It was clear to me that this was not the time for makeshift measures. Our only choice was to initiate radical reform, including the downsizing of our photography-related businesses. Had we delayed by just another year or two, we would have been right in the middle of the devastating financial downturn in the fall of 2008 and the company might not have been able to survive this double punch.

At the same time, while moving ahead with structural reforms in our photography-related businesses, we boldly invested in those fields that we felt showed future promise. By developing entirely new ventures, we could fill the vacuum left by the loss of our core business.

After assuming the position of president, I used each and every opportunity to stress the fact that this marked a new start for the company, a "Second Foundation," if you will—much more on that later. For an entire decade it is fair to say that Fujifilm braved high waves, kept on course, and managed to traverse a turbulent sea.

Thanks to this series of reforms, the company was literally reborn. In 2012 our former perennial rival, Kodak, filed for Chapter 11 bankruptcy. Fujifilm, however, escaped being swallowed up by the transformations of the times and continues to evolve even today.

How was it possible for Fujifilm, in the throes of the most critical crisis, to carry out such radical reforms? What decisions did I make from my position of leadership, and how were they carried out?

Many people have asked me these questions, including media from around the world. Some suggested that I give my answers in the form of a book. The present volume is the result.

Today, many Japanese corporations feel they are operating under a stifling cloud, with the future uncertain. I hear of firms in every industry whose traditional foundations have been shaken to the core as they confront the watershed moments of our age. I have here taken up my pen in the hope that an account of how Fujifilm overcame the most serious crisis in its history will be of some service to the reader.

> *What happened at Fujifilm?*
>
> *What did Fujifilm do?*
>
> *What do businesses need from their current leaders, and from the leaders of the future?*

The following is the story of the managerial reforms I undertook, and, in Part Two, my thoughts on the theme of leadership.

PART ONE
Fighting for Fujifilm

THE CORE BUSINESS VANISHES

WHAT HAPPENED AT FUJIFILM?

The transformation was coming, and it would be profound. It would disrupt our industry in the way that word processing disrupted typewriting, music CDs displaced records, and email dispatched the handwritten letter. Growing and accelerating like a snowball swelling to an avalanche, it was aimed right at us. In fact, our name put us in the crosshairs.

The transformation was digital photography, and we were Fujifilm.

The Coming Crisis

When I was appointed president of Fujifilm in 2000, sales of the company's primary product line of photosensitive materials, notably our color photographic film, recognized around the world in its signature green boxes, had just about peaked. The following year, Fujifilm sales finally overtook those of the Eastman Kodak Company, the industry giant, something we had been striving to do since our foundation. It was a corporate mission embraced by everyone at Fujifilm, and achieving our goal was a source of great pride.

The sales gap between the two companies as I signed on with Fujifilm at the beginning of the 1960s was immense and daunting. Kodak had more than ten times our revenues, and it took Fujifilm nearly forty years to catch up to our mighty monolithic rival. But we did, and our success was mighty in itself. By then, our market share in Japan alone was close to an overwhelming seventy percent.

The peak of the industry coincided with our own peak—Fujifilm's emergence as the worldwide market leader in photographic film. Yet, in an often unpredictable business world, a peak always conceals a treacherous valley. It soon became clear that we would preside in our leadership over a contracting market. Digital cameras were already spreading like wildfire, basically eliminating the need for film. After

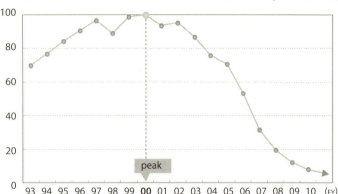

the industry's sales peak in 2000, the film market began shrinking very slowly, then picked up speed, and finally plunged at the rate of twenty or thirty percent a year. Ten years later, worldwide demand for photographic film had fallen to less than a tenth of what it had been.

This is how a battle began that I couldn't afford to lose, a battle that I would throw myself into body and soul.

My task was to save the company.

Technology-Oriented Fujifilm

Fujifilm today is engaged in such a wide variety of businesses that it's difficult to summarize everything we do. But you can say one thing with certainty: this company is technology-oriented.

Fujifilm was founded in 1934 with the mission to undertake the domestic production of photographic film. The need for high-quality water and clean air led Fujifilm to establish its factory and headquarters at the foot of Mt. Fuji in Minami Ashigara City, Kanagawa, Japan. Many product lines were developed at this location: photographic film and photographic paper, X-ray film, graphic arts film, PS (pre-sensitized) plates, magnetic tape, related equipment, IT products, and more.

Japanese companies around this time assumed that establishing a new company or creating a new industry required importing technology from abroad. But at Fujifilm we decided to develop our own technology, and this meant the company had to clear a number of immense hurdles. The postwar transition of film from black and white to color was particularly difficult. Not only Fujifilm, but Kodak and other companies throughout the world were struggling with the development of this new technology.

There were at one time thirty or forty producers of monochrome photo and X-ray film in existence globally, among them some of the world's most renowned chemical companies. But with the advent of color film, many of these companies were confronted by an insurmountable wall—the control of material and quality at a level of precision they could not meet—and they were forced to withdraw from the business.

After DuPont's and 3M's departure, the only

firms left in the color film business were America's Eastman Kodak, Japan's Fujifilm and Konica, and Europe's Agfa-Gevaert. The history of photographic film manufacturing thereafter rode on the rivalry among these four companies in quality control and technology.

The Kodak Giant

Kodak was head and shoulders above all the others in the manufacture of photographic film when I joined Fujifilm in 1963. At the time, Fujifilm had total sales of ¥27 billion, a figure dwarfed by Kodak's nearly ¥400 billion.[*]

Then, of course, the difference between all Japanese companies and their American rivals—think Toyota and General Motors—was much the same story. And the difference was not just in sales. Kodak's technology was also far ahead of Fujifilm's. It was true as well of brand recognition and financials. The gap between these two rivals was simply enormous.

The year 1963 was only eighteen years after Japan's defeat in World War II. Most Japanese at the time thought that American companies were invinci-

---

[*] Because of fluctuations in the yen-dollar rate, cost figures are given in yen throughout this book to preserve their relative values. In 1963 dollars, ¥27 billion = $75 million. ¥400 billion = $1.1 billion.

ble, that it was fruitless to even try to compete. At one point I heard something from a senior colleague that made my jaw drop. Large amounts of silver and gelatin are needed to manufacture photographic film, and my colleague told me that Kodak had its own silver mine and ran its own cattle ranch to secure a supply of high-quality gelatin from cattle bones! We might just as well have been taking a knife to a gunfight.

This was the situation when Fujifilm began its relentless pursuit of Kodak. Our corporate goal was to surpass Kodak—eclipse our Rochester, N.Y. rival— because it was the colossus of the industry. Despite the huge hurdles, we would be the giant slayer.

Fujifilm's technology caught up with Kodak's in the late 1970s, and by the 1980s we firmly believed we had technically surpassed Kodak in nearly all varieties of film. From the 1980s into the 1990s, we waged a persistent struggle with Kodak for world market share. And when we took the lead, it was not at all a Pyrrhic victory, even if worldwide film sales almost immediately began to fall.

It was, instead, the beginning of an unprecedented challenge.

First Attempts at Diversification

My first job at Fujifilm was in the Corporate Planning Division. As a young, untested executive, the only assignment I was given consisted of forecasting demand for photographic film. Desk work didn't suit me, I found—I was just too ready for action, and I soon asked for a transfer to sales as a better channel for my pent-up energies.

Fujifilm at that time had a Consumer Photo Products Division, which was in charge of photography-related matters, and an Industrial Materials & Products Division, which handled photographic technology and other newly developed technologies for industrial use. I was assigned to the latter.

The company had set up the Industrial Materials & Products Division in 1961, a couple of years before I joined. But Fujifilm had already been involved in related areas, such as the development of X-ray film, since shortly after its founding in 1934, and had a history of leveraging its photographic technology to other products and markets to give its business activities a broader scope.

Photo film is made of a finely tuned combination of various basic technologies. For example, in color photography, an original color image is reproduced essentially by mixing the three primary color materials used in printing: cyan (C), magenta (M), and yellow (Y)—these are complementary to the three primary colors of light: red (R), green (G) and blue (B). But if the balance between them is even a little bit off, the photographic image of a white color shirt, for example, may have a red or bluish tinge when it is printed. Precise control of materials is essential to maintain a proper color balance in the reproduced image.

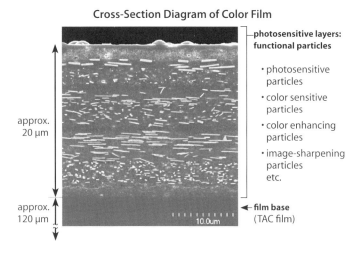

Look at a cross section of color film and you will see that on a clear base film there are twenty evenly coated layers, each sensitive to the three primary colors of light—red, blue, and green—that are perceived by the human eye. These overlapping twenty layers are no more than twenty microns in thickness. Thanks to high-precision coating technology, these light-sensitive layers are able to capture the faintest illumination, amplify it, produce color, and record an image. The reproduction of color in such thin layers is an amazing technical feat that was accomplished long before the semiconductor industry managed to squeeze ever more transistors on a chip. In fact Fujifilm today makes many products for that industry, too.

Various advanced technologies are required to

manufacture photographic film. In addition to film formation and high-precision coating, there are grain formation, functional polymer, nano dispersion, functional molecules, and redox control (manipulating the oxidation of a molecule). Inherent in all these is very precise quality control.

It was only natural that Fujifilm would consider whether some of these technologies could be used in other ways. For example, instead of a film base, coatings of photosensitive material can be applied to aluminum plates, which are then exposed and developed so that lettering and graphic areas can be coated in ink to produce a master plate for printing. This is the so-called PS plate used in offset printing. By expanding on the technology of photographic film and replacing silver halide with magnetic powder as a base coating, it becomes possible to develop and produce audiotape, videotape, and computer memory tape.

At the time I joined Fujifilm, this type of diversification was already proceeding quickly. Understandably so, since it was during this period, too, that the import of photosensitive materials into Japan was being liberalized and custom duties were being lowered. This meant that Kodak, with its overwhelming technological lead and sales volume ten times that of Fujifilm, would soon be entering the Japanese market.

Preparing for the inevitable, Fujifilm threw itself wholeheartedly into developing photographic film products that could compete on an international level. We lowered costs by rationalizing our manufacturing processes. We also made plans to enter new fields of enterprise, leading to the establishment of the Industrial Materials & Products Division, to which I was assigned.

At the time, PS plates and associated products were new on the market, so when I visited a printing company it was not unusual for the printer to say, "Fujifilm, huh? What brings *you* here?" And I would say, "Well, it happens we have this product called a PS plate ..." I spent practically every day cold-calling on potential clients.

The biggest moneymaker for the company at the time, by far, was still the area of photography products, which was handled by the Consumer Photo Products Division. But it was the graphic-arts-related business of the Industrial Materials & Products Division that would later make a huge contribution to a sudden surge in our sales—and fortunes.

Overseas Development and Foreign Encroachment

In 1976, Fujifilm was the first company in the world to create high-speed color negative film with its development of F-II 400. This was an epoch-defining event that essentially meant Fujifilm had surpassed Kodak in technology.

Riding the momentum, Fujifilm launched an aggressive worldwide campaign in the 1980s, proclaiming its technologically sophisticated products under the slogan "Challenging the World with Our Technological Prowess." Fujifilm built up its international manufacturing base and sales network and began to compete with Kodak in every corner of the world, establishing itself as a truly global enterprise. In 1984, Fujifilm became an official sponsor of the Olympic Games in Los Angeles, Kodak's own U.S. backyard, and in other ways continued to bolster its marketing activities. We built a factory in the U.S. in 1988, and continued to grow our share of the American market.

This reversal of long-standing American and Japanese commercial positions was apparent in a number of other industries in the 1980s. Many Japanese enterprises caught up with and technologically surpassed their American counterparts and assumed superior competitive positions. In 1985, with much of the world feeling the threat of Japan's growing industrial strength, the G5 nations (France, Germany, the U.S., the U.K., and Japan) gathered at the Plaza Hotel in New York City and drew up the Plaza Accord, which called for an adjustment in the currency exchange rate. This marked the first of many times, continuing to the present day, that Japan and Japanese industry have had to shoulder the burden and competition of a strong yen.

But this type of international pressure on Japan was not confined to the exchange rate. At about the same time the U.S. government began to apply successively more political pressure, through such collaborations as the U.S.-Japan Semiconductor Agreement and the U.S.-Japan Automotive Consultative Group, to force Japan to increase imports into its own domestic market.

Many of these political tactics were directed right at Fujifilm's core business: photosensitive materials. In 1995, Kodak filed a petition with the Office of the United States Trade Representative arguing that Japan's alleged exclusionary practices with regard to photographic film and paper had closed the Japanese market to competition to favor Fujifilm.

Fujifilm, under the direction of then-President Masayuki Muneyuki, produced a detailed rebuttal based on objective fact and categorically refuted the charges. In 1998, the World Trade Organization announced a sweeping rejection of Kodak's complaints, giving Fujifilm the victory in this trade skirmish. This case was unusual for the time, showing that Japan could compete directly with foreign countries—and win. The victory was highly regarded in Europe and elsewhere, and Fujifilm's presence rose in sales and stature in markets throughout the world.

The Approaching Digital Era

Just as Fujifilm and Kodak were competing fiercely against one another, a bigger threat was approaching—one that would require a fundamental change in both organizations. That threat was the digital age and the radical transformation in the photography market that accompanied it.

By the beginning of the 1980s, industry watchers were already predicting that silver-based photosensitive materials would one day be an endangered species. In Fujifilm's principal imaging fields—photography, printing and medical—the first signs of digitalization had already begun to appear.

In photography, companies were already developing cameras that used a photoelectric element instead of film to capture an image. They were known as "electronic" cameras then, or what we call digital cameras today. Trial products were already being tested by electronics makers and photo-related companies.

In the printing field, an Israeli company in 1979 announced a computer-based device for plate-making and information processing called a Response System. The technology for digital X-ray imaging for medical diagnostics also appeared around this time. Fujifilm in 1981 developed the first digital method for diagnostic X-ray imaging, using a special photosensitive plate in place of X-ray film. The technology led to the creation of new Fujifilm products, and in 1983 Fuji Computed Radiography (FCR) was brought to market. The resulting system is still the world standard thirty years later.

The digital age drew steadily nearer and nearer to Fujifilm's core, but the biggest shock for me, as a member of the Industrial Materials & Products Division and dealing with graphic-arts-related matters, was the digitalization of plate-making.

Until then, the plate-making process used in printing newspapers, magazines, books, and so much more had made extensive use of film. The cover of a full-color magazine, for example, required that the typography, the photos, and the illustrations all be produced separately and then assembled by hand— a complex job that might call for dozens of pieces of film. For us in Industrial Materials & Products, plate-making from film was a very big, and very important, market.

Then, with digitalization, it became possible for a computer to create a full-color plate, assembling all its components via software and outputting the finished page on a single piece of film.

I was manager of the sales section at the time, and I couldn't help but worry about the future: "As digital printing becomes established," I thought, "it's possible that film will no longer be needed. This is big."

Allaying my fears a bit, I figured that when the digital age did come, it would arrive in stages. It seemed unlikely to me that a technological revolution of such magnitude would come in one fell swoop. By my reckoning at the time, it would take another thirty years for digital cameras to catch up to analog cameras in terms of resolution and sensitivity.

Still, I felt strongly that the battles to be fought

in the coming digital age would be of a different order entirely, and extremely tough. Because the film business called for the highest level of technical expertise, the number of competing companies had already declined by natural selection until there were just four of us remaining. There was price competition, of course, but it was still possible to sell at a reasonable price point and sustain a profitable business if you maintained high technical standards, top-grade image quality, and products superior to your rivals'.

The digital age, I sensed, would be different. It would be a world in which Fujifilm's proprietary technical expertise—the photography technology built up over the years, including high-precision coating of chemicals on film—would no longer be relevant.

Digitalization, after all, is a kind of standardization. The black box is small, and there is little room for differentiating one product from another through technology. I believed that the digital age would not be a battle between rival technologies, but an inescapable price war.

What's more, that price war would be intense, because the technological barriers to entering the field of battle would have been lowered, and the competitor group of four would crumble. The profits enjoyed by the photography industry would be a thing of the past.

"This is going to be a survival game of the most brutal kind," I thought.

Three Strategies for the Digital Age

My sense of crisis was naturally shared by Fujifilm's upper management. We discussed—and debated—what Fujifilm should do in the face of this digital wave, and eventually summoned up three strategies:

Develop Original Digital Technology

The digital age was coming no matter what, and it should be regarded not as a crisis, but as an opportunity. So the first strategy was that Fujifilm should develop its own original digital technology and become a pioneer of the digital photography era.

We vigorously took on the challenge, and Fujifilm in fact developed the world's first fully digital camera. Research began in the 1970s, and 1988 saw the creation of the Fujifilm DS-1P, with fully digitized input and output. The first compact camera with a resolution comparable to silver halide film was also produced by Fujifilm, in 1998: the FinePix700, with a resolution of 1.5 megapixels.

Indispensable to our taking the lead in digital cameras was developing the CCD image sensor, among other innovations, in-house. Thus, when the demand for digital cameras took off at the end of the 1990s Fujifilm was a front-runner with a thirty percent share of the market.

By successively taking on the key technologies of the digital era, such as imaging software and digitalization of printing plates, as focuses of research,

World's first fully digital camera: DS-1P.

FinePix700.

as well as through large corporate acquisitions overseas—about which more is to come—Fujifilm intrepidly prepared itself for the digital age.

Extend the Life of Photosensitive Materials

The second strategy was to extend the life of our photosensitive materials business by raising analog image quality to a level beyond digital reach.

The fact was that photosensitive materials using a silver halide base still had a good deal of room for improvement. We launched research on raising the film's level of light sensitivity so that a flash was unnecessary. Also, the grain was made even smaller, increasing resolution. The goal was to produce an image from photo film that was far superior to anything from digital technology. And even today, many photographers say that they like texture of film and sometimes use a film camera.

Develop New Businesses

The third strategy was developing new businesses that were neither digital nor based on photosensitive materials. Our traditional market for photosensitive materials was generally limited to a few key competitors. But it was easy to imagine a future in which digital became mainstream and price competition turned fierce, making it impossible to sustain previous levels of sales and profit. That future was almost certainly coming. We would have to create new businesses to make up the difference.

In the mid-1980s, we began research on inkjet and optical disk technologies, fields peripheral to our traditional imaging business. At the same time, we diversified even further, launching research on cancer drugs in cooperation with Dr. Susumu Tonegawa, recipient of the 1987 Nobel Prize in Physiology or Medicine.

Pharmaceuticals, because they employ many similar technologies, are considered next-door neighbors to the speciality chemicals used in photography.

From the 1980s onward, I took every opportunity to urge our then-president to consider undertaking new kinds of business. I was still in sales at the Industrial Materials & Products Division, in particular graphic-arts-related materials, and I couldn't help but feel a sense of crisis as I noticed the slow but steady change in a market that was moving toward digitalization, though it was different from the rapid market transformation seen later in photographic film. Even though I was only a section manager in those days, I spoke personally with the president more than a couple of times to champion the changes that were needed.

"In the coming digital age the technological assets built up by Fujifilm over the years won't necessarily be viable," I put forward. "It will be a whole new world, a battle of different technologies. The competition will be fierce, and it won't be profitable. To prepare for that day, we need to immediately start something new and big."

Unfortunately, most of these new businesses ended in failure. The principal reason for this, ironically, was that the highly profitable core business of photographic film was still growing.

The 1980s were a period of economic affluence in Japan, and a period in which the global demand for photographic film was still surging as the worldwide photography market continued to expand. Fujifilm by then had overtaken Kodak technologically, and our profits were rising rapidly throughout the world.

Then, in 1986, we gave ourselves another reason not to buck the status quo. The "film with lens" camera, the Fujifilm Quick Snap, went on sale. A digital camera doesn't need film, but Quick Snap was film that didn't need a camera. Essentially a roll of film in a box with a shutter and lens on the front, it was a lifestyle-changing product that led to further growth in the demand for film. Fujifilm at the time was consistently producing an annual operating profit of slightly more than ¥100 billion.[*] Our operating profit at its peak rose to ¥180 billion a year.

Hints of the future threat to photographic film profits had already emerged, but when you looked at the present, sales of film were stunningly good. This performance ultimately undermined the development of new businesses. After all, new businesses were not immediately lucrative, and even if a promising new

---

* In 1986, ¥100 billion = $617 million.

segment should emerge, it could never match the profitability of photographic film.

Most of us believed that, so it wasn't a surprise when some concluded that "Photographic film is still expanding and producing a profit, why would we want to try something new?" Fujifilm, as a result of what Americans would call this "horse-and-buggy" mindset, abandoned inkjet technology and optical disks, and sold off its newly established pharmaceutical company.

Brave Enough to Recognize Reality?

Meanwhile the progress of digitalization stalled. Even after digital cameras had been available for some years, their image quality remained inferior to that of film, and aside from some journalists who really needed to transfer photos electronically on deadline, digital cameras didn't gather momentum.

Yet the global photographic film market kept growing. Within Fujifilm itself the mood was optimistic: "Digital will never catch up with analog in terms of resolution." "Not everything can become digital." "Photo film is good for another thirty years." Or so many people thought.

In the end, given the situation, top management decided to forgo investment in new businesses. It was a matter of "When you already have a highly profitable business, if it ain't broke, don't fix it."

Conditions were good; there was no denying that. But to anyone looking at the present coolly and unflinchingly, it should have been apparent that this situation was not going to last forever. To prepare for what was coming, we should have taken more risks and made plans for the future.

These conclusions didn't just apply to Fujifilm. The same can be said for any industry, and for any company, at any time. No matter how good business is, you have to foresee and prepare for a coming crisis. Looking directly at reality, you have to recognize what is happening at the moment, as well as what is going to happen in the future. You have to read the situation, understand it, think about it, and decide what needs to be done. This is what management is all about. If your reading is wrong, and the times change, your company will eventually face a crisis.

I was promoted from section manager to department head in 1985, and I continued to urge top management to prepare for what was to come: "We should invest in the future," I encouraged them. "We should find new sources of income."

It's not that top management lacked any sense of crisis. When I became a director and general manager of the Industrial Products Marketing Division in 1995, then-President Minoru Onishi gave me the special assignment of looking into the effect of digitalization on Fujifilm's business and technology. Four people on my staff immediately began the research with me, and we produced a thirty-page report.

Five years later, when I was made president, I couldn't help but feel that that special assignment had been a kind of message, an expression of hope and expectation, because what we had predicted was where we were going.

The Crisis Arrives

I became president of Fujifilm in 2000. By then, nearly all manufacturers' camera lines, aside from professional models, had begun the transition to digital. But our photography product line, which included film for general use and color photographic paper, still accounted for sixty percent of Fujifilm's sales and two-thirds of our profit that year.

It was also in 2000 that photographic film hit its peak. In 2001 the global demand for color film suddenly plunged. The speed and impact of its descent exceeded everyone's expectations—and their nightmares. The photographic film market shrank at the rate of twenty to thirty percent a year. A market that had accounted for sixty percent of our sales and two-thirds of our profit had disappeared in the twinkling of an eye. In five short years our core business, which for many years produced huge sales and comparable profits, had fallen into red ink. The digital wave of the future had become a reality in just half a decade.

And what about the digital camera, where Fujifilm had been a forerunner in technological development? It was good that we had made the investment. With the expansion of digital demand, sales continued favorably, and Fujifilm grabbed a top share of the world market.

But my biggest worry lay ahead: competition based on pricing. Camera prices soon dropped—as with all consumer technology, it seems—and continued to drop at the rate of fifteen percent per year. Digital cameras simply could not make up for the loss of income from declining sales of photographic film.

My fears had come to pass. If we didn't do something now, there'd be hell to pay. In 2003, I became CEO. I had tough decisions to make, and changes to push through. Half measures wouldn't do it. Steeling myself for the task, I was determined to execute drastic organizational reform at Fujifilm.

This company had long contributed to society by producing high-quality products as a leader in the fields of photography and imaging. But now, unless something was done, it would cease to exist. The technology and other business assets so carefully developed over the years would all come to nothing. Somehow, Fujifilm had to be kept alive as an enterprise that meant something to society. The lives of more than seventy thousand employees worldwide, and their families, were on the line.

SIDEBAR

Digital Minilabs: A Godsend for Photo Stores

Although Fujifilm recognized early on that the demand for photographic film would shrink drastically because of digitalization, we were also aware that future demand for photo prints from digital cameras would likely increase. This led us to the idea of photo stores serving as centers for printing out digital data. To deliver the best product for customers, we decided to have the photo stores use color photographic paper, which was far superior to other types of digital printing paper in cost competitiveness and quality. This would have the additional benefit of allowing the photo stores to remain in business longer.

A digital minilab called Frontier, developed solely by Fujifilm, made it possible—in fact, created the market. Frontier was installed in photo stores throughout the world to print out digital data—which is what digital photos in reality are. This compact printer processor helped the photo stores gain a new source of income, making up for the revenue they were losing as the business

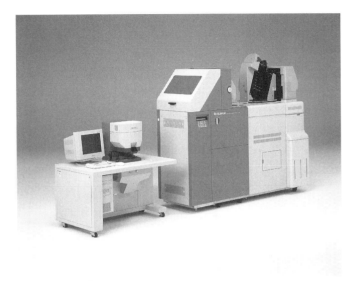

Digital minilab Frontier.

of developing photographic film rapidly dwindled.

The photo stores were invaluable business partners for Fujifilm. They had previously maintained themselves by developing film and making prints—a lot of it our film and on our paper— and we strongly felt that we should do something to shelter them from the pains of a world going digital.

Frontier enabled photo stores to produce prints that were far more beautiful, of better quality, and longer lasting than those from home printers. Printed digital photos emerged as a big market, and riding the wave, photo stores gained a renewed life.

Offsetting all the losses incurred by the shrinkage of the photographic film market was impossible, of course. Many photo stores closed. But thanks to the minilab, many others managed to stay in business. For many a photo store the minilab was a godsend, and I am still of the opinion that it had social significance—the Frontier minilab met customers' needs for real photos they could share in person with friends, and at family gatherings.

The minilab spread to supermarkets in the United States and then to the wider world; it broadened the scale of our business, helping Fujifilm gain and maintain its top global share.

Of course, perennial rival Eastman Kodak was pursuing the same line of business, too. Whereas Fujifilm had developed the system in-house from the beginning, Kodak had relied on OEMs—original equipment manufacturers. Fujifilm, in this segment of the business, never lost a battle to Kodak.

The minilab would later have a positive effect on our Digital Imaging Division as well.

For example, it can automatically adjust the lighting of faces in pictures taken where there's bright light behind the subject. Fujifilm's original image-processing technology would later play a vital role in the development of facial recognition technology in digital cameras.

A SECOND FOUNDATION

FUJIFILM'S CHALLENGES AND REFORMS: THE WHOLE STORY

Global demand for photographic film peaked in 2000, the year I became president, and then suddenly went into free fall. More than ever I felt a sense of crisis, and that we had to find some new business. In 2003, when I was appointed CEO, I began a series of radical reforms. The following ten years marked the birth of a new Fujifilm, its "Second Foundation."

My Return from Europe

In June 1996, four years before I became Fujifilm's president, then-President Onishi sent me to Europe.

"I'd like you work on the outside," he said.

"On the outside?" I asked. "You mean in a subsidiary?"

"No. I mean Europe." He wanted me to become president of Fujifilm Europe.

Three times in the past there had been talk of an overseas transfer, but I had declined on every occasion. At the time I was caring for my aging mother, and I just couldn't leave her behind and go abroad. But after my mother passed away, the situation was different. I had become a general manager and a director, and there was no longer any reason not to go. It was my time.

In the fall of 1996 I took up my duties in Düsseldorf, Germany, and began supervising the company's European strategy. Those four years in Europe were a kind of rehearsal for the reforms I would later be compelled to initiate in Japan.

Almost thirty years had passed since Fujifilm's move into Europe, and its share of the market there stood at about twenty-five percent. Kodak held forty percent, and many felt that Fujifilm had to resign itself to holding second place.

Fujifilm's biggest issue in Europe was that it

had become used to being the perennial runner-up.

"There's no way we can compete with Kodak."

"We should be satisfied with twenty-five percent. There's no need to knock ourselves out."

This kind of thinking pervaded the company. The local staff as well as newly arrived personnel from Japan were content with the status quo; it was the easiest path to take.

But I wasn't happy with either our position or the attitude that accepted it. To me, it was obvious that we could, and should, do better. Through its high quality, extensive product lineup, and brand appeal, Fujifilm had already surpassed Kodak to become No. 1 in Japan. Even when priced at a premium to Kodak, Fujifilm's products still sold strongly.

But the situation in Europe was completely different. Our products were priced a good ten percent lower than Kodak, in some stores, seeking to appeal to bargain hunters rather than those looking for quality. Fujifilm was in no way inferior, so why did we have to eat Kodak's dust? Our real appeal wasn't being fully appreciated, I thought. Fujifilm's standing in Europe should be much, much higher.

I read the European staff the riot act.

"Just look at our quality and our lineup. It's great," I chided them. "Why should we be satisfied with being No. 2?"

Just building a fire under them wasn't enough, of course. I immediately put together a list of ideas for reform and confronted the staff with it: rethinking the brand strategy, reassessing pricing, introducing new products. I gave them all the reasons we could be No. 1 and gave them the means to get there. After that, I told them, "It's up to you."

I also ordered sales subsidiaries and distributors in each country to come up with ideas for new sales strategies. And as the company's top salesman, I went out to client meetings myself to close the most important deals.

In Fujifilm's three decades in Europe, the Europeans had apparently never seen a Japanese person like me. I could tell they were a little taken aback. I learned later they referred to me as a "samurai." I took it as compliment.

Back in Japan, I'd become known as someone who could handle a difficult situation. When I was appointed general manager of recording media products, the price war for videotape was at its fiercest. And when Fujifilm bought out a manufacturer of digital plate imaging devices, they sent me to head up the Graphic Arts Systems Division and confront the issues in managing it.

So in 2000, just when we were making good inroads into Kodak's stronghold in Europe, I was called back to Japan. This time, as the digital floodwaters were rising, they were asking me to be president, and there was not a moment to lose in carrying out fundamental structural reforms.

"They're calling on me again when there's trouble" was my first thought, so I didn't feel that I was being asked to do something entirely new. The scale of the job as president would be larger, but it was the work I knew I'd long been getting ready for. I had always exercised leadership with the company's interests foremost in mind.

As the No. 1 man in Europe, I had developed new products and restructured the organization. Now, the company's core photographic film market was shrinking at a spectacular rate, and the situation was critical. Fujifilm had good management resources, first-rate technology, a sound financial footing, a reputable brand, and excellence in its diverse workforce. If all these assets could be effectively combined into a successful strategy and applied, I was sure that something could be done to save the day. The whole of Fujifilm was depending on my managerial skills to make it happen.

It was a daunting realization, but I was gripped by a strong sense of mission. "Maybe I was brought into this world to overcome this crisis," I thought at the time. The hair stood up on the back of my neck.

Not Just to Survive, But to Thrive as a First-Rate Enterprise

I was appointed president and COO in June of 2000, but I didn't have the final word in controlling the company. I still had a boss, the CEO, and of course there was always the board of directors. I worked diligently to manage Fujifilm's health, planning the vital reforms that would assure its long-term well-being after I became CEO in 2003.

In carrying out reform, only one thing occupied my mind: making Fujifilm a company that could—and would—play a leading role throughout the twenty-first century.

If the goal were only simple survival, many things could be done. We could get rid of unprofitable businesses. The company would still be financially sound, after all. For many companies this would undoubtedly have been a viable option. But I had absolutely no intention of taking that path.

Fujifilm had, until then, been one of the leading companies in the photographic products industry and had continually produced big profits. I wanted to make sure it stayed that way into and through the next century. Figuring out how to do it was my job as CEO.

My first task was to draw up a plan to make the reforms we needed. With Fujifilm's core photographic film market crumbling, it was my job to determine our future direction, the type of company Fujifilm should be, and a practical program for achieving those ends—and finally to communicate all this to the company's employees, whose motivation was essential in making the plan work.

On February 5, 2004, we announced VISION 75, a medium-term management plan in honor of Fujifilm's 75th anniversary. It would extend to the fiscal

Medium-Term Management Plan "Vision 75"

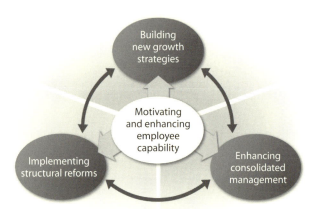

year ending March 2010—a blueprint for implementing fundamental reforms and changing the structure of the company, with the vision of "saving Fujifilm from disaster and ensuring its viability as a leading company with sales of ¥2 or ¥3 trillion a year."

The three policies incorporated in VISION 75 were "implementing structural reforms," "building new growth strategies," and "enhancing consolidated management." These objectives could only be realized by employees who were highly motivated and possessed superior skills.

As they watched the core business of photographic film disintegrate before their eyes, many employees must have been concerned about the future of the company. I had to show them what was going to be achieved at Fujifilm from here on out. Along with announcing VISION 75, I rallied them with the reality of what it meant to do nothing:

"In our present situation, we are Toyota if *cars* were to disappear. We are Nippon Steel if *steel* were no longer needed. As the demand for *photographic film* steadily disappears, that is precisely our situation. We have no choice but to confront it, and to confront it head on."

Reorganizing and Consolidating

The photographic film business is built atop a giant industry infrastructure. Fujifilm had large-scale factories in Japan, the United States, and the Netherlands, as well as photofinishing labs in one hundred fifty locations throughout the world. Maintaining facilities on this scale was extremely costly.

Once sales started to drop, they dropped without stopping, and deficits led to more deficits.

If we decided to give up photographic film, then all these costs could be eliminated without a second thought. But giving up film was not something I was going to do.

Determined not to abandon a market where our technology and reputation still promised a future, but faced with continued falling demand for photographic film year after year, I had to reorganize the business to ensure a stable flow of profit. This would necessarily involve some serious downsizing to create a smaller, more flexible business that was in keeping with current demand.

VISION 75's first principle, "implementing structural reforms," guided us in 2006 as we set about reorganizing the photographic film business, including Fujifilm's global, large-scale manufacturing plants and sales organizations, research centers, and photo-finishing labs.

For Fujifilm's in-house New Year's message that year, I wrote a piece called "Moving Toward Real Reform," which laid out the stark reality:

> *Comparing the interim results for 2004 and 2005, imaging net profits fell by nearly ¥10 billion, producing a ¥5 billion loss.* We can't wait any longer—now is the time to ruthlessly cut production and manufacturing facilities, as well as the sales-related operations that became bloated during the company's heyday, and spare only what is essential.*
>
> *Now is the time to maximize efficiency to the utmost, to fight as if our lives depended on it, to challenge unknown emerging markets, and ultimately to save the photographic film business.*

---

* In 2005, ¥10 billion = $84 million.

Pushing Relentless Reform, But with Consideration for All

When VISION 75 was announced in 2004, I was fully aware that fundamental structural reforms were needed, but I thought that if manufacturing were reorganized, sales reformed, and purchasing and procurement rethought, we would not have to cut back personnel.

In fact, we did a detailed simulation of the effect that reducing costs would have after transitioning to the new structure, and found that it could indeed be done.

But what we could not account for in our projections was the speed of the digital onslaught. The photographic film market had shrunk much faster than we expected, and about two years later we came to the realization that it wasn't going to work out after all. I knew we had to downsize.

The announcement was made in January 2006.

At the time the Fujifilm Group had some fifteen thousand global employees in film-related areas. We reduced that number by five thousand, saving as many jobs as we could by transferring people to different lines of work. In addition, the sales licenses of four large-scale underperforming film dealers in Japan were called back and combined into one in-house function.

Of course I didn't want to fire anyone or cancel the licenses of special dealerships. Nobody likes downsizing. No one is going to simply say, "Sure, let's just do it." Things aren't that easy.

But if the company went under, there would be nothing left—lives and careers and a business built by outstanding work all gone up in a puff of smoke.

I just had to grit my teeth and make the decision. A CEO—really any top-level manager—is responsible for thinking about the future, twenty or thirty years ahead, or even more, to ensure that the company survives and thrives. What has to be done has to be done, with determination and resolution. That's the job of a leader. Of course, it has to be done with care and consideration for employees and partners who've labored valiantly for the company's success.

Downsizing is far from inexpensive, but fortunately Fujifilm was financially well-off. We forgave the debts of our primary distributors, bought back their sales licenses, and made it possible for them to pay severance to their employees. These dealers had been long-time business partners, comrades fighting shoulder to shoulder, and it was simply unacceptable for their owners and employees to suddenly lose everything because Fujifilm's contract had been cancelled.

Over the next year and a half the downsizing continued.

The total cost of our restructuring came to over ¥200 billion.[*] Though the film market continued to

---

[*] In 2007, ¥200 billion = $1.8 billion.

shrink at an ever-faster rate, our efforts to substantially and quickly downsize the film business while cutting back fixed costs were a success, in both timing and scale. If we hadn't made the hard decision when we did, there would have been hell to pay.

Preserving the Culture of Photography

Fujifilm announced its plans for reform to the world, including the intent to downsize its photographic film business, on January 31, 2006. The media pounced on the news, focusing on the reduction of personnel in headlines that screamed: "Fuji Photo to Cut 5,000 Jobs."

Just ten days earlier, Konica Minolta had announced its complete withdrawal from the camera and photographic film business. Given the advent of the digital camera, as well as the shrinkage of the photographic film market by twenty to thirty percent a year, anyone seeing such headlines might be forgiven for thinking, "Aha, now it's Fujifilm's turn!"

But I never thought, either then or now, of withdrawing from the photographic film business. On the day Konica Minolta made its announcement, I sent out a press release stating that Fujifilm would honor its photographic film legacy.

Under the headline "Concerning Fujifilm's Photo Film Business," I declared that "we will continue our photography business centered on silver halide film and strive for the further enhancement of the culture of photography." A few days later this statement made page one in the *Asahi Shimbun*'s well-known op-ed column *Tensei Jingo*.

Why did I so prominently proclaim that Fujifilm would continue with photographic film? Some investors asked me, "Why don't you just give up?" Some even warned me, "This is going to drag your business down."

But a company is more than just profit and loss. Photography is an extremely important aspect of human culture. Photographs are splendid reminders of those wonderful times in our lives. They can capture and preserve moments of joy, glorious moments, memorable moments spent with one's family. When we take them out and look at them, photographs bring back special times from the past and rekindle the feelings we felt, just as we experienced them, however long ago it may have been.

Fujifilm's mission, I believed then and I believe today, is to preserve and sustain the culture of photography. It's not a question of making or losing money. No doubt the film market would continue to shrink, we knew, but that was precisely why I thought Fujifilm had to continue giving its support to the miracle of photography.

FUJITAC and the Market Growth of Liquid Crystal Television

In the metaphor-filled world of the business press, Fujifilm's moves toward structural reform were once described as "putting on the brakes while stepping on the gas."

On the one hand, we were carrying out decisive cuts in the photographic film business, while on the other, we were investing heavily in new businesses we thought had a promising future. One example was our investment in polarizing plate protective film, an essential ingredient in the manufacture of liquid crystal panels, which make TVs, computers, cell phones and so many modern devices come alive before our eyes.

A polarizing plate allows the passage of light traveling in a specific direction, and what protects the plate is the protective film. We developed a film, called FUJITAC, that is made of natural materials, is highly transparent, and possesses superb optical characteristics. Since the 1980s, FUJITAC has been used in liquid crystal displays (LCDs) for calculators, and as they switched to flat-panel displays, in personal computers and televisions as well. In fact, the flat panels of LCD TVs have two polarizing plates, and each plate needs two protective films. It has grown to be a huge global business.

The polarizer protective film market was essentially led by Fujifilm and one other firm. Fujifilm's global share stood at more than seventy percent. But at the time, it was still uncertain whether liquid crystal or plasma displays would become the standard for television. Plasmas don't use the same technology, and don't require our protective film. I put my money on liquid crystal, and even before the market began to expand, I built a new facility (Fujifilm Kyushu) in Kikuyo-machi, Kumamoto, Japan, to serve as the production center for FUJITAC. It was a bold—if risky—decision to bolster up the supply side. Total investment in the plant rose to over ¥150 billion.

The market for LCD televisions did undergo explosive growth, and Fujifilm was right there as a premier manufacturer and supplier of polarizer protective film. I believe that our ability to meet the demand played a huge role in the spread of LCD TVs.

Polarizer protective film is now used not only in televisions, but also in smart phones, and this business segment has turned out to be a pillar of profit for Fujifilm. What was only ¥2 billion in sales at the beginning of 1990 became ¥200 billion twenty years later, easily making up for the losses incurred as the photographic film industry so severely contracted.*

* * *

I have an indelible memory connected with FUJITAC film. Toward the end of the 1960s, when

---

* In 1990, ¥2 billion = $15 million.

Fujifilm Kyushu.

FUJITAC production line.

I had just been transferred to sales in the Industrial Materials & Products Division, I was in charge of the FUJITAC product. At that time the economy was experiencing a temporary downturn, and FUJITAC's sales were in a slump. Photographic film, on the other hand, was doing quite well. The idea was floated to shut down what was then the tiny FUJITAC business.

I pleaded with my boss. "Don't shut it down. I'll dig up something new for them."

Camping out at the office, I worked on strategy day and night, and then brainstormed latent markets with an engineer, coming up with ideas such as electronic billboards that proved successful.

The results were good, and FUJITAC's life was extended, eventually incorporating the polarizer protective film business. If I had done nothing and had simply acquiesced to FUJITAC's demise, that business might never have come into existence. It was well worth the fight.

Looking back, that was a real turning point for me. I clearly remember thinking, "If I don't live up to the company's expectations and keep FUJITAC alive, my presence at Fujifilm is basically meaningless." It compelled me to work frantically to develop new markets and keep the business afloat.

Everybody experiences a number of turning points in life. For me, this was one of those times. These are make-or-break situations, and no matter what, you must not buckle, you must not back down. When I became president and the materials used in flat-panel displays, centering on FUJITAC film, had become one of Fujifilm's core businesses, I felt as if fate had played a hand.

Needs and Technology: Searching for New Business

The second policy proposed by VISION 75 was "building new growth strategies." Structural reform may be a critical need, but if reform is the only goal the company will end up frozen, in a state of contracted equilibrium, making it impossible to remain an industry leader.

The digital world is a world of ruthless price-cutting. Even though Fujifilm had succeeded in producing a digital camera and had come to terms with digitalization, those milestones were not enough to capture back the former profitability of the film market. We had to create a highly profitable core business in its place.

Companies can't embark on new initiatives with fuzzy thinking like, "Oh, it looks promising. Let's give it a try." It's corporate suicide. We had to ask where could we make use of our technological assets, our business resources. Without knowing the answers, we couldn't expect to successfully launch a new venture. We needed to ascertain Fujifilm's particular strengths, view them in an organized fashion, to confirm their essential value.

Quadrant Map

	Current		New	
New	laser endoscope	next-generation inkjet printers		regenerative medicine
	super dense computer backup tape	EA-ECO toner	ultrasound diagnostic devices	pharmaceutical products
	SYNAPSE medical imaging and information management system	production system printers		
	high-quality digital cameras	multifunction copiers	semiconductor materials	cosmetics, supplements
Current	digital cameras (compact)	copiers, multifunction devices	conducting films	heat-resistant film
	computer backup tape	PS plates, CTP	CIGS semiconductor material for solar cells	solar cell backsheets
	Fuji Computed Radiography (FCR)	optical lenses		
	color paper	X-ray film	film for LCD	
	photo film		mobile phone plastic lenses	

TECHNOLOGIES (vertical axis) / **MARKETS** (horizontal axis: Current — New)

Even before becoming CEO, I had ordered the head of R&D to take inventory of Fujifilm's technical stock, its technological "seeds." I then told him to compare these seeds with the demands of international markets and review the results. After nearly a year and a half, we had a quadrant map to help blaze new trails.

Current technology and new technology were on the vertical axis, and current markets and future markets were on the horizontal. The map inspired us to ask ourselves the following questions:

> *Given current technology, are there further applications for current markets?*
>
> *With new technology, are there additional applications for current markets?*
>
> *Given current technology, are there new applications for new markets?*
>
> *With new technology, are there new applications for new markets?*

With this tool in hand, we came to a renewed appreciation of Fujifilm's technological capabilities and its potential for meeting new market needs. We saw that Fujifilm's technology could be adapted for emerging markets such as pharmaceuticals, cosmetics, and highly functional materials, and we identified six priority business areas that could act as driving forces. We began their implementation by pursuing two approaches: strengthening existing businesses showing current growth, and creating new businesses for future growth.

Not Just Success, But Long-term Success

We selected the business areas that would become

the core of our future growth strategy. But we didn't choose them simply on the basis of whether we could produce a product. We didn't just focus on whether we could outperform the competition.

The question was, rather, did Fujifilm have the ability not just to win, but to win and continue to win? Did we have the basic, fundamental technological skills to carry out a world-beating strategy? That was our gold standard.

We spent considerable time evaluating each technology and each product strategy, as well as how they fit with our overall business strategy. Our decisions had to take everything into account. And as president, I had to make the final call. If I misread the prospects for our new ventures, all of Fujifilm's reforms would come to nothing.

And since it was our company, and our future, we relied on ourselves. Discussions concerning new businesses were kept in-house. I continually told my staff, "Don't rely on outside consultants. Think for yourselves!"

Of course, there are times when you have to seek outside experts for advice, for those areas where maybe you don't have the internal competence. But in the end outside advice is nothing more than outside advice. To ask strangers what you should do about your own company, especially when you're betting your future, to me is out of the question.

Any top-level manager who has to rely on outside advice to make a final decision should quit immediately. Maybe that seems harsh, but it's truly what I believe.

Time and again we discussed not only possible new ventures, but also investments in ongoing businesses, such as the FUJITAC polarizer protective films. FUJITAC seemed to have great potential, though it wasn't yet clear whether liquid crystal or plasma was going to be the display of the future. Still, we had to make a choice. After gathering as much information as we could, we came down in favor of liquid crystal, leading to our huge investment in the Kyushu factory. I'm happy to say we chose wisely. This factory met the needs of the coming explosion in flat-panel televisions and showed remarkable growth.

Finally we chose six business areas to focus on:

Digital Imaging

The star player in this segment is the digital camera. Its three key components are the lens, the sensor, and the processor, and the technology for all three was created in-house at Fujifilm. Doing our own development work is one of Fujifilm's strengths. Putting this capability to good use, the company is now devoting considerable resources to its X-Series and other high-end cameras with significant added value.

Other products in this area include printouts from digital cameras, smart phones, and photo books. Fujifilm is very competitive with its offerings for all of them.

Optical Devices

The main products in this area are television lenses, security camera lenses, mobile phone camera lenses and so forth. I'm happy to say that Fujifilm has great technological expertise in these as well. Even in TV camera lenses, which require the most sophisticated optics technology available, Fujifilm has captured more than fifty percent of the global market. We saw we had the strengths in our in-house, high-precision optical technology and mass-production skills to make this another ongoing focus for Fujifilm. Because we made the decision, Fujifilm is now a huge player in value-added, high-precision, high-quality image resolution and image quality in smart phone modules and related products.

Highly Functional Materials

Polarizer protective film is the undisputed king of our highly functional materials business. At present, Fujifilm owns over seventy percent of the global market. Recently we've begun selling touch-panel sensor film, solar cell backsheets, and other highly functional films with a great future.

Graphic Systems

This is our graphic-arts-related sector, including devices and materials for digital printing, and we expect this area to see extraordinary global demand continue and grow. Fujifilm has expanded four factories around the world to produce CTP (computer-to-plate) materials for the master plates used in offset printing, today the standard for most mass-produced publications. At the same time, we're increasing our efforts in digital printing. By using our in-house ink-jet technology, we have started introducing our own epoch-making, next-generation digital printers into the market.

Document Solutions

This business is mainly handled by our subsidiary Fuji Xerox, in which we are the majority partner with the American business process and document management company Xerox Corp. In addition to office copiers, multifunction devices, and printers, Fuji Xerox handles graphic systems for the digital printing market and provides solutions to enhance customers' operating efficiency through the improvement of their office documentation and business workflows.

Healthcare

Fujifilm has long supplied state-of-the-art X-ray film and diagnostic imaging systems to the healthcare industry. In addition to bringing these products to an even higher level of quality and usefulness for our customers, Fujifilm has now moved into the area of functional skin care cosmetics and dietary supplements, adding the prevention side of healthcare to our traditional diagnostic area. And moving recently into pharmaceuticals, Fujifilm is also taking a role in patient treatment.

Healthcare's Importance in the Twenty-first Century

It would be a mistake to underestimate the importance of healthcare, both to the modern world in general, and Fujifilm in particular. No matter where they live or what their age or circumstance, people value their health. So it's no surprise that healthcare will be an increasingly vital business in the twenty-first century. Medical technology may be advancing at a remarkable rate, but many seemingly unsolvable challenges remain. These challenges are opportunities for Fujifilm's leading-edge medical technology to offer solutions and make a real difference in healthcare throughout the world.

In 1936, not long after its founding, Fujifilm got its start in the medical field with X-ray film, and in 1971 then-subsidiary Fujinon began selling endoscopes. In 1983, the world's first digital X-ray diagnostic imaging system was made available by Fuji Computed Radiography (FCR), which has become the world's de facto standard. Even today, although the basic patent has expired, FCR still holds a larger share of the global market than any competing product.

In 1999 we introduced SYNAPSE, a medical-use picture archiving and communications systems (PACS) that allows digital images to be saved on a hospital server and called up whenever and wherever needed. Today, SYNAPSE is installed at some four thousand medical facilities throughout the world.

Fujifilm in 2012 bought SonoSite, a major American maker of ultrasound diagnostic devices. SonoSite has a large share of the portable ultrasound market, and with its purchase Fujifilm entered this industry sector in earnest, aiming to make it a pillar of growth in the medical systems area.

Centered primarily on imaging, Fujifilm has long made contributions to medical diagnosis, but as the technology of healthcare has grown, we have expanded into it, hoping to make healthcare a long-term mainstay of our growth. We plan to continue our expansion in the preventive fields of cosmetics and dietary supplements, adding to our established expertise in diagnosis, and to keep expanding in pharmaceuticals as well. By covering prevention, diagnosis, and treatment, Fujifilm has set its sights on becoming a comprehensive healthcare company.

The Rationale behind Cosmetics

Why cosmetics? "Makeup" made by a photo film company may strike you as odd, but the two fields have an amazing amount in common. The chief ingredient in photographic film is gelatin, which is derived from collagen. Human skin is seventy percent collagen, to which it owes its sheen and elasticity. In the course of developing and continually improving its photographic film, Fujifilm gained nearly eighty years of experience in research on collagen. We became experts in the science of human skin while working on something completely different.

For example, there's the process of oxidation. Oxidation is connected both to the aging of human skin and to the cause of photos fading over time. Discovering how photos deteriorate, and what can be added to the mix to prevent that from happening, was and in fact still is a specialty at Fujifilm. Taking stock of our expertise here gave us the idea of taking our long experience in creating anti-oxidation technology for preventing color photos from fading and applying it to the development of anti-aging cosmetics.

And there's also our competency in emulsions—one material suspended in another. Photographic film is coated with organic compounds sensitive to yellow, magenta, and cyan suspended in gelatin, and there are many experts at Fujifilm who know how to micronize these compounds, dissolve them, and create emulsions.

Emulsions are required in three major fields: cosmetics, food products, and photographic film. So it's no wonder that for a long time many Fujifilm engineers have taken an interest in makeup. When I first heard the idea of making cosmetics as a new business venture, my ears perked right up. Using Fujifilm's technology, I realized, we could produce something that no other company could.

Today, Fujifilm makes a brand of functional cosmetics called ASTALIFT with the anti-oxidizing compound astaxanthin, a natural ingredient extracted

Functional cosmetic ASTALIFT.

from vegetable matter. As it happens, astaxanthin is oil soluble and very hard to handle since it does not dissolve in water. This is where nanotechnology and our experience with emulsions come into play. Using our own nanotechnology, created during work on photographic film, Fujifilm made it possible to dissolve usually indissoluble matter in water, enabling it to be absorbed efficiently by the skin where it is needed. It was an innovation that a pure cosmetics company could have never brought to the table.

ASTALIFT was launched in 2007, with Japanese celebrities Seiko Matsuda and Miyuki Nakajima as spokesmodels. Our line of cosmetics has been very well received.

Cosmetics still account for a small portion of our total sales, but I expect that Fujifilm's technology will continue to lead us to very special products that provide real value to the company—not to mention to their satisfied, younger-looking users. R&D is underway, and we expect to be actively promoting the sales of cosmetics not only in Japan, but soon throughout the world.

A Full-Scale Entry into Pharmaceuticals

In March 2008 Fujifilm acquired mid-level drug development company Toyama Chemical in a takeover bid and made a full-scale entry into the pharmaceutical products market. Toyama Chemical is an outstanding company that has developed a number of innovative technologies ("seeds"), and we knew that by combining our technologies we could create unique new products with added value.

After becoming a consolidated subsidiary within Fujifilm, Toyama Chemical worked in cooperation with our manufacturing engineering division to improve its manufacturing processes and reform its facilities, leading to a quick rise in profitability. Collaboration between the two companies moved ahead in other areas as well. We carried out joint research on new pharmaceuticals, while the clinical trials of Fujifilm's new drug FF-10501 (for patients with relapsed or refractory myelodysplastic syndrome—a type of blood/bone marrow cancer) were left in the hands of Toyama Chemical, which had the right experience.

Toward a Comprehensive Healthcare Company: Prevention, Diagnosis, Treatment

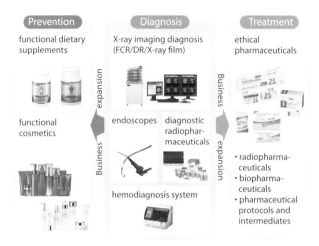

By mutually exploiting each other's assets—expertise in pharmaceuticals and in the components of photo film—we created a new synergy that few would have thought possible.

Delving further into the healthcare segment, Fujifilm also bought a radiopharmaceuticals sales and development company (now called Fujifilm RI Pharma), a venture business that began at the University of Tokyo for researching antibody drugs (presently Perseus Proteomics), and a biopharmaceuticals contract manufacturing organization (now FUJIFILM Diosynth Biotechnologies), among others.

In a joint venture with Kyowa Hakko Kirin, Fujifilm established Fujifilm Kyowa Kirin Biologics to produce biosimilars, which extend the choices of highly complex drugs made by manipulating living organisms. And to further broaden its healthcare business activities, Fujifilm has invested in Japan Tissue Engineering, which is engaged in the development and production of regenerative medical products such as Autologous Cultured Epidermis (JACE) and Autologous Cultured Cartilage (JACC), both of which use a patient's own cells to help "grow" new biological material for treatment and repair.

Developing effective medicine involves finding relevant chemical compounds, but it also calls for technological innovation in how drugs are absorbed into the human body. That, today, is one of the most competitive venues in the pharmaceutical business. The competition has just begun, and no one company leads the pack as of yet. If Fujifilm's nanotechnology can be applied to pharmaceutical products, their absorption rate will be enhanced and the medicine delivered to the affected area more quickly. A photo film company might just become a prime innovator in medicine.

The development of macromolecular pharmaceutical products using biotechnology is accelerating, both by Fujifilm and others. Quality and cost are key. The manufacturing engineering skills and quality control technology it has accumulated over the years give Fujifilm the ability to develop and produce competitive products of high quality and reliability. In the next few years, by 2018, Fujifilm aims to triple its net sales in healthcare to ¥1 trillion.* That's how highly Fujifilm estimates its potential in this area.

Mergers and Acquisitions Provide a Head Start

To get new business ventures underway, Fujifilm has made active use of mergers and acquisitions (M&A). Speed is of the essence in today's business world, and starting from scratch is many times just not efficient; there wouldn't be enough time to build a presence in the marketplace. By acquiring companies that have already left the starting gate and combining their assets with Fujifilm's expertise, we can get new products to market quickly and easily.

One particular event made me realize the possibilities of M&A. This occurred in 2001, a little before the announcement of VISION 75, with the consolidation of Fuji Xerox.

At the time, Fuji Xerox was a fifty-fifty joint-venture partnership between Fujifilm and Xerox—an American company with operations centered in Rochester, N.Y. Perhaps because we had invested only fifty percent, we ended up having little say in the management of the company.

Further, collaboration between the two companies was not making much headway—and Fujifilm was doing little but collecting dividends. When I

---

* In 2014, ¥1 trillion = $9.6 billion.

became a director in 1995, this was the first issue that I brought up at the board meeting.

In 2001 Fujifilm bought an additional twenty-five percent of Fuji Xerox shares and raised its stake to seventy-five percent, making the JV a consolidated subsidiary. As expected, there was a great deal of overlap between Fuji Xerox's dominant document solutions business, including office multifunction devices and printers, and Fujifilm's long-established printing materials and photographic business. The synergy produced by combining them was strong, and to me a clear indication of the benefits in such an acquisition.

M&A has enabled us to move immediately into many promising areas, such as inkjet printing, medical devices, and pharmaceuticals. In the inkjet field, Fujifilm acquired the top company in the world for print heads. It also acquired companies focused on medical IT systems, software technology, and, as I discussed in the previous section, biopharmaceuticals—in all about forty companies. Altogether, we spent close to ¥700 billion on mergers and acquisitions.,

As the head of Fujifilm, I don't like wasting money; I want to hold on to what we've earned. I don't want to borrow; I want us to be financially independent. But the vital point is building for the future. This is make-or-break time. When we commit, our chances of success must be good; and once we have embarked on a new venture, we have to be ready to fight to the end. It's essential to make investments

Principal Strategic Mergers and Acquisitions

Month/Year	Sector	Description
2/2006	Graphic systems	Acquired Avecia Inkjet Limited, a leading maker of dyes and inks for inkjet printers
7/2006	Graphic systems	Acquired Dimatix, Inc., an American maker of industrial inkjet printer heads
10/2006	Medical systems	Acquired all shares of Daiichi Radioisotope Laboratories, Ltd.
12/2006	Medical systems	Acquired Problem Solving Concepts, Inc., an American provider of medical imaging information systems for cardiology
1/2008	Imaging	Acquired IP Labs GmbH, a German online photo service system developer
3/2008	Pharmaceuticals	Acquired Toyama Chemical Co., Ltd., in a takeover bid and made it a consolidated subsidiary
12/2008	Medical systems	Acquired Empiric Systems, LLC, an American manufacturer of radiology information systems as a wholly owned subsidiary
12/2008	Medical, imaging	Acquired Russian distributor Fujifilm-Ru
3/2010	Medical systems	Acquired Brazilian distributor NDT Comercial Ltda.
10/2010	Pharmaceuticals	Formed a capital alliance with Japan Tissue Engineering, a developer and distributor of regenerative medicine
2/2011	Pharmaceuticals	Acquired two leading companies in contract manufacturing of biopharmaceuticals from the American company Merck & Co., Inc.
3/2012	Medical systems	Acquired Sonosite, Inc., a major American maker of diagnostic ultrasound devices
8/2012	Document solutions	Acquired the business process outsourcing (BPO) unit of Salmat Limited, Australia's leading business services provider

for the future and take that calculated risk, even if it means some current sacrifice.

As to where we invest, the point is to produce something via a new synergy that is different from what is being made elsewhere. Our investment is aimed at producing a unique innovation that other companies do not have. This strategic synergy is Fujifilm's overarching selection standard for M&A. Just increasing sales by bringing in a new operation alone means nothing. To win at a new business, you have to produce something truly excellent through the combined efforts of the two companies. This is what we at Fujifilm consider most vital in M&A.

Creating a New Center for Interdisciplinary Research

When we were deciding what new businesses to enter, one non-negotiable condition was that if we were going to produce a winner, Fujifilm's technological prowess had to be brought into play. Our research capacity was at the core of our strength, and VISION 75 called for the restructuring of research and development as one of the company's most important goals.

Fujifilm possessed a variety of exceptional technologies in fields such as chemistry, electronics, mechatronics, optics and software, and among Fujifilm employees were a great many outstanding researchers. I'm proud to say there are few companies aside from Fujifilm that can boast of having both so many advanced technologies and so many talented research personnel.

Yet it wasn't as though there were no problems. Our research system was predicated on the photo business, and our research centers were scattered here and there, but always near our factories, according to function. The Ashigara Research Laboratories, for example, did research on photographic film and photographic paper, and the Fujinomiya Research Laboratories on non-silver halide aspects. The labs were sited close to Fujifilm factories and did research on what those respective factories were producing. Along with being widely dispersed, they were mainly doing research on subjects related to photographic film, and that market was rapidly shrinking. When film was in its heyday, this system might have worked well enough, but times had definitely changed.

While technology in general has been advancing rapidly, very few issues today can be addressed and answered by one technology alone. More and more, problems are best solved by combining the appropriate, though often disparate, technologies. This led us to the notion of an interdisciplinary facility that would develop core technologies for creating new businesses and products, a place where researchers from a wide variety of fields across the company could come together to do advanced work. We decided to build a new research center where the scientists and technical experts could collaborate and thrive.

For the design, I gave the concept to the top R&D managers and had them work out the actual plans for the facility. The building was erected in Kaisei-machi, Japan, on a huge 7.3-acre lot and cost ¥46 billion.* With multiple stories, the R&D center would be even bigger: total floor space of 635,000 square feet, twice the area of the land it sits on. It was a colossal structure.

This new center was named the Fujifilm Advanced Research Laboratories, a research and development center where nearly a thousand highly specialized scientists, engineers, and technicians could gather in one spot and conduct cutting-edge research.

The center operated under three guiding principles:

> INTELLECTUAL FUSION: *To fuse the thought processes and knowledge of specialists from different fields.*
>
> INNOVATION: *To create disruptive innovations/technologies and new value networks.*
>
> VALUE CREATION: *To provide society with new customer value.*

We paid meticulous attention to the research environment in order to execute these mandates

---

* In 2006, ¥46 billion = $392 million.

Fujifilm Advanced Research Laboratories.

Interior of Fujifilm Advanced Research Laboratories.

effectively. The researchers primarily work in a spacious open room except when they're in the individual laboratories. There are no walls, and it is not unusual for the individuals in adjacent islands to be doing entirely unrelated work. The conference rooms are enclosed in glass. Whiteboards are everywhere in partitionless meeting spaces, with open discussions taking place wherever they arise. The library is called the Knowledge Café, from the notion that the library is not a place to study but a place to communicate ideas.

These concepts were suggested by the researchers themselves and have resulted in an environment where the knowledge and thinking of a great variety of engineers and scientists can be fused efficiently together.

Ongoing R&D Investment of ¥200 Billion a Year

No matter how unpredictable the company's financial situation was as we implemented the reforms, I refused to cut back on R&D investment, including building the Fujifilm Advanced Research Laboratories. If anything, I increased spending. Even in the toughest of times, we managed somehow to put together ¥200 billion a year for R&D.

In the eleven years from becoming president in 2000 until I was named Chairman and CEO in 2012,

Fujifilm's R&D investment reached a total of ¥2 trillion.*

Top-level managers of publicly traded companies are perpetually questioned about their success in the marketplace, and one essential measure of that success is profitability. Often, by cutting back on R&D they can easily add three or four percent to their profit margin as a percentage of sales. So managers are continually tempted to cut R&D expenditures and have to struggle against its financially seductive lure.

And yes, it would have been easy to constrain Fujifilm's R&D expenditures and improve profitability when I became president and began to transform the business. But it was one thing I refused to do. Even while cutting other costs, I did not want to reduce investment in research and development; it was essential for future growth. If anything, I wanted to expand R&D.

A firm must possess a certain culture, an ingrained way of thinking, to continue to create innovative products. This is precisely why there's a need for investment in the future. It comes down to the question of what a corporation should do to secure its future growth. For Fujifilm, the ten years following 2003 were the years when investment was needed to plant the seeds. At its peak, our ratio of R&D spending to net sales reached eight percent.

In one sense, management efficiency—the urge

---

* ¥2 trillion is roughly $19 billion averaged over this period.

to achieve the highest profitability possible—was sacrificed during this period. Put another way, there are times in the management of a company when efficiency has to take a back seat.

Middle- and long-term investment does not produce immediate results; it takes time for the seeds to bear fruit. During that period, short-term management efficiency undoubtedly suffers. But technology investments will pay off in the end and become a mainstay of the company—as long as your research goals are on target.

I have always believed in investing in the future. Even in my younger days, I urged upper management many times to think about the need for investment. If you don't invest when you should, what is the use of investment at all? Particularly in the 1980s, when the company was beginning to expand abroad and realize huge profits, it should have risked investing in core technologies that would have provided a backbone for future growth.

That kind of bold investment wasn't carried out, which inevitably led to crisis in the days to come.

Transition to a Holding Company

The third basic policy to VISION 75 was "enhancing consolidated management."

Fujifilm owned some excellent subsidiaries, such as Fuji Xerox and Fujinon, the optical lens business for mobile phones and broadcast TV cameras, but so far there had been little collaboration between them and hence not much synergy. We had to bring ourselves closer together.

So we created FUJIFILM Holdings to further consolidate management, and Fujifilm and Fuji Xerox became its subsidiaries. The idea was to unify management and eliminate the silos, ensure a valuable mutual exchange of ideas and creativity among the engineers, and benefit from the resulting synergies.

In particular, I thought that the synergy with Fuji Xerox could be raised to a much higher level. Both the market and the technology in graphic arts, copying, and photography had already become cross-border enterprises. There should be much more collaboration on a company level as well.

In 2006 the company's structure transitioned to a holding company system, and the headquarters of both Fujifilm and Fuji Xerox were relocated to the Tokyo Midtown complex. I set up a committee with members from both companies to discuss what technological contributions each could make and how they could work together. We saw a number of proposals emerge from this discussion, and many of them were implemented, such as the idea for a next-generation inkjet printer. Fujifilm would make the inkjet heads and Fuji Xerox would be in charge of setting up the system.

In addition to the holding company, we introduced a cash-management system to administer group funds, thus controlling finances more effectively, and we transitioned operations held in common, such as personnel and general affairs, to a shared company called Fujifilm Business Expert. As a result, we raised business efficiency and lowered costs for the whole group.

A New Name: "Fujifilm"

When Fujifilm transitioned to a holding company, we also made another big decision—to change our name as well.

The company name was originally Fuji Photo Film, and for the long period when sixty percent of our net sales were produced by photography-related products, that name and the well-known trade name Fujifilm reflected the nature of the company's core business. But as the film market shrank and our diversification increased, Fujifilm had become more of a conglomerate. We still produced photographic film, but it was no longer a core product, and our name struck some as an anachronism—potentially even a liability.

So we invited ideas for the new name. Since we would become a holding company, some suggested that the name be radically changed. We spent some time discussing made-up names like "Fujix." So many corporate names had been invented out of thin air—

but to my mind, there was something wrong with all of the proposed new identities.

What seemed most important to me was that the name Fujifilm was already widely known throughout the world. The company's logo and billboards, with their unmistakable green chromatic theme, were global presences. Fujifilm was a world brand.

When I consulted an expert in the field I was told that it would take ¥200 billion to achieve that level of recognition today. With that much inherent equity, it made simple sense to continue using the brand we had firmly established over the years.

And although photographic film was now only one of our many businesses, it remained the bedrock of Fujifilm's technology. If you explained we were still engaged in photo film this way, as a foundation for all that we are, the name Fujifilm continued to make total sense.

Our Best Performance in History— Then the Global Collapse of 2008

The efforts we made in the transformation and expansion of our business—including our structural reform, large-scale investment in existing businesses, funding new ventures, and the organizational changes in transitioning to a holding company—had borne fruit for Fujifilm in a big way. In the fiscal year ending March 31, 2008, Fujifilm recorded a record-breaking ¥2.8468 trillion in sales and an unprecedented ¥207.3 billion in profits.* I enjoyed positive, reinforcing thoughts like "Our structural reform is steadily progressing" and "If you do what is right, good results will follow."

But deep down my honest feeling was, "So far, so good." There were still reforms to be carried out, though from here on, I felt, it was just a matter of pressing forward in much the same fashion. I even dared think to myself, "My work is done."

No sooner had those words formed in my mind than we were struck by a calamity that was not of our making, that the world didn't see coming, and that nearly brought the global economy to a standstill.

What has come to be known in Japan as the "Lehman shock"—named for the crisis set off by Wall Street giant Lehman Brothers' bankruptcy filing in September 2008—was followed by a worldwide depression. Around the globe, markets for nearly every product line from every industry began to shrink at an extraordinary rate.

Like a few others whose positions open windows into the global financial system, I had become aware of disturbing signs nearly a year earlier. But it was only in the autumn of 2008 with the economic downturn that I fully realized that we were entering a great depression, and that "the unthinkable"

---

* In 2008, ¥2.8 trillion = $30.7 billion; ¥207.3 billion = $2.3 billion.

was happening. One day I was looking at the sales figures for one of our core divisions when my jaw dropped.

"What the hell is this?" I asked. The division had achieved only seventeen percent of its monthly sales target.

I immediately knew this was something that should not be. My long experience in sales told me that a seventeen percent drop would be understandable, but an achievement rate of a mere seventeen percent was astonishing—that is, astonishingly awful. Even the closing rate fell below fifty percent.

If the same thing were to happen in every division, the company would teeter on the brink of collapse. The fact that something appalling had occurred couldn't be denied. Reality was staring me in the face from the monthly report and I had to accept it for what it was. Deep in shock, I wondered, "What's to be done?"

A Second Companywide Restructuring

It's a question that needs a good—and expeditious—answer: When faced with a sudden catastrophe, what is top management to do?

The first thing is to determine how far-reaching the effects are, and how long-lasting. Even if the picture should turn ugly in the short term, you can usually deal with it smoothly if you know that it's temporary. However, if it seems the situation is not going to settle down, it's time for forceful measures.

I pondered what the effect of the global financial crisis would be, and how long it would last. How much would the economy shrink, especially in our most important markets? At first it was extremely difficult to predict where things were headed, but around the end of the year things began to come clear. It was going to take some time before this staggering economy would heal.

From the beginning of human history, economic activity has followed an upward trend, and I believe that trend will basically continue (with the usual, manageable fluctuations). Why? Because human beings have a natural desire to improve their lives so that today is better than yesterday, and tomorrow is better than today. As long as humans have that desire, the economy will have to expand to keep up with it.

However, that's the long-term view—the trend line from thirty thousand feet. There will always be recessions—some of them serious recessions—and sometimes recovery will be slow. Unfortunately, with the recession that began in 2008 there was little hope for a quick recovery. This was not a simple fluctuation but a global economic crisis. My analysis said the market would probably make a small comeback, and then shrink to about seventy-five to eighty percent of what it had been. After that, it would begin to gradually expand. That's how I not very optimistically saw the situation.

Once the impact and duration of a catastrophe such as 2008 have been determined, it's the job of the top manager to bring the company into line with the new situation—that is, to make the necessary, though often unpleasant, adjustments to the corporate structure. If the market is growing smaller, it will prove difficult if not impossible to turn a profit by doing business as usual.

Operating on the premise that Fujifilm's market would shrink by twenty percent, it was my job to remake the company so that it would still make a profit.

Fortunately, the previous year we had had a record-breaking performance, and Fujifilm was financially robust. The structural reform of the photo film-based business undertaken in 2004 had achieved its initial successes at this point—had it not, that failure in combination with the global financial crisis would almost certainly have brought us to our knees. The task now was to revisit fundamental structural reform throughout Fujifilm to get us through the crisis—and make sure we were in a good position to go beyond.

To create the type of company that could produce a profit under even the most dire conditions, we began restructuring all of Fujifilm's divisions and reducing the cost of selling, as well as general and administrative expenses—that so-important-to-manage SG&A. At the same time, we pushed ahead with a growth strategy for our priority businesses.

I felt it was important to convey precisely and unmistakably to each and every employee the meaning and purpose of the restructuring—this was hardly business as usual—so I held a "town hall" meeting to speak to them directly. But I didn't stop there. I also created the opportunity for managers to talk to smaller groups of ten or so employees and convey the same message in a more informal fashion.

From the fiscal year ending March 2010 to the fiscal year ending March 2011 we moved with determination. Reforms were boldly—if not always happily—put into effect. Inefficient facilities, organizations, and work routines all received a thorough review, and five thousand back-office positions were trimmed from the administrative and R&D divisions and elsewhere.

Fujifilm thus cut its fixed costs by over ¥100 billion annually. In the fiscal year ending March 2011 this led to a substantial gain in operating profit over the previous year; net sales showed a slight increase.

One Misfortune Follows Another: The Strong Yen

After the downturn of 2008 Fujifilm undertook a complete restructuring of all its divisions and remade itself into a company that could be viable under the most uncompromising conditions. Even if the market shrank twenty percent, we had transformed ourselves into a muscular entity that could still turn a profit.

This was the second time Fujifilm had stepped back from the brink by enacting drastic reforms, and it had been a success.

But it wasn't the end of our trials. It was just then that yet another crisis struck: the strong yen. Before the recession, the yen stood at ¥110 or ¥120 to the dollar, making Japanese products very attractive to America and the world business community that does so much of its trade in dollars. But following the fall 2008 downturn, the yen-dollar exchange rate appreciated thirty to forty percent, and finally stabilized at about ¥80 per dollar.

For Japanese companies competing around the world, this was the worst imaginable situation. We cut costs at Fujifilm even further and tried every available means to combat the strong yen, but when your home currency continues appreciating at the rate of thirty or forty percent a year, you can't make a profit, try as you may.

More than one member of our top management in China and South Korea told me, "I don't know how you can continue operations with the yen so strong." It was a wonder to them that we could stay in business at all.

Around this time, in 2009, the Democratic Party of Japan took over the reins of government, managing things ineptly until 2012. Happily, it was followed by the Liberal Democratic Party and the Abe administration. Since then, the exchange rate and the strong yen have been correcting. Yet, the damage suffered by Japanese companies during those four or five years cannot be easily healed. We took it right in the pocketbook.

The most grievous impact of the strong yen has been on the enthusiasm of many Japanese top-level managers and workers. The manufacturing sector was particularly hard hit, and the spirits of those workers watching their companies become so uncompetitive around the world greatly suffered.

Japan still has steps to take to fix the overly appreciated yen and get the economy back on its feet. At the same time, the Japanese people need to assume a more optimistic outlook. It won't be easy, but we'll get there.

Back on the Path to Growth

There've been a number of recent events, one after the other it seems, that have had an enormous impact on the Japanese economy, in addition to the strong yen. The economic crisis in Europe was one, and that's not over yet. There have also been natural disasters, such as the 2011 Great East Japan Earthquake and tsunami, as well as the flooding in Thailand.

The scars left by these events still haven't healed. And we know we'll be faced with these kinds of trials and tribulations again and again in the years to come. It doesn't make me fear the future. In fact, bring it on! I can handle it.

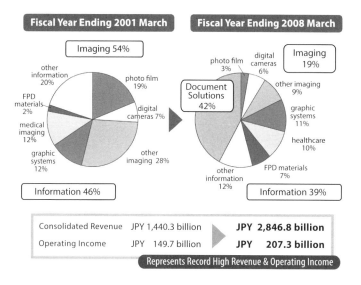

A Diversified Business Portfolio: Growth from Adding New and Expanding Existing Business Areas

This is all to say that management is an endless series of individual battles. To not end up a loser, the company has to be made more muscular, more robust. Our combat readiness has to be sharpened; our level of performance has to be raised. This is the thought that as Chairman and CEO of Fujifilm continually occupies my mind.

Since assuming the position of president fifteen years ago, it has been my relentless aim to see Fujifilm survive and develop as a leading company through organizational restructuring and reform: namely, by transforming the company away from the photographic-film-centered enterprise it was since its founding; by diversification; and by making technology the core of each of our business segments. All of these were intended to lead to a "Second Foundation"—the birth of a new Fujifilm.

This Second Foundation, however, is still a work in progress. While it is true that we rode out the digital storm, in my estimation we are only about seventy-five percent of the way to finishing the job. To be sure, we've achieved some degree of success. Compare our recent sales breakdowns with those of the past and the difference becomes clear. Thanks to the managerial reforms that we undertook, the company has been completely transformed, and is now well on its way to making the most of that transformation.

In the ten years since I became CEO of Fujifilm, it's been my mission and mandate to carry out radical corporate reforms to come to grips with the critical loss of our core business. And I've done it. I worked with staff to draw up the plans, and then I handed down the final decisions. That's been my job, too.

But just drafting plans and deciding to go with them is not enough. They have to be implemented, put into practice, and for that you need the overall strength of the company—the strength of each division, department, and individual worker. It is when management takes firm steps in the right direction, and when people are motivated to follow that lead, that a company can be turned around.

If there is one thing to be said about me and my management style, it is that I have always made firm decisions. If I make a hundred decisions, I am absolutely sure, every time, that this one is the right one. And once I decide something, I stubbornly stick to it, even if that means dragging others along with me.

It is a little early to let our guard down. We have survived one crisis, that is true, but the company is like a patient who has survived major surgery. The new business areas are not fully competitive, and profitability is not what it should be. We are in a transitional period, in a valley between two mountains, moving toward the next period of growth. After my more than a decade at the helm, storm-battered Fujifilm has been set on a true course and assured a safe passage.

During my years at Fujifilm I have received much from the company, learned much, grown, and matured. And this in turn has enabled me at last to pay back some of what I have received. I feel honored to have had this opportunity, and blessed as CEO for all that I have gained.

What's Next?—After the Trunk and Limbs Come the Branches and Leaves

Fujifilm has, with the establishment of new business ventures, completed the first stage of the structural reform that accompanied the sudden shrinkage of its core photographic film market. We have, in effect, established a tree with a trunk and substantial limbs.

But to become a fully formed tree, we need to grow branches and leaves as well. We're entering the second stage of reform under difficult conditions. We have to foster the further development and growth of each individual business line.

This second stage requires a thorough review of our sales and marketing capabilities, our speed and effectiveness at R&D, the competitiveness of our product pricing, and the value creation by our administrative divisions.

We have to ask ourselves some important questions:

Are our products cost competitive?

Are they selling at a cost proportionate to their performance?

Are they capturing market share in line with their functionality?

Is our R&D competitive in terms of speed and efficiency?

Is there no room for improvement in the effectiveness and efficiency of our administrative departments?

All these areas have to be scrutinized, and if they're lacking, brought quickly up to snuff.

In 2012, I became Chairman and CEO of Fujifilm, and Shigehiro Nakajima was named President and COO, marking the beginning of a new hierarchy. I would decide the general direction of the company and make final decisions, but it would fall to President Nakajima to manage everyday affairs and spearhead reforms. Our aim was to raise our capabilities throughout the Fujifilm enterprise.

President Nakajima today is playing a central role in promoting our "G-up" initiative, whose goal is to empower each and every employee. If every worker in manufacturing, sales, R&D, and administration creates more value in his or her areas of competence, then Fujifilm will remain fully competitive, even in the worst of economic conditions.

I firmly believe we can guarantee it.

Another important task for the second stage is to add in the pieces of the puzzle that were put on the back burner during the first stage. There is always more that can be done. For example, in the fields of cosmetics, diagnostic equipment, and biopharmaceuticals, Fujifilm needs to develop products with new functionality and to thrust ourselves more assertively into Asian and Western markets. The acceleration of change and bolstering of these existing businesses will be major themes of our ongoing evolution.

Creating a Company That Can Create Change

When we at Fujifilm turn our eyes to the wider world, we seem to see no end of problems—the stagnation of the European economy, the economic slowdown in developing countries, the rising costs of raw materials, and more. Competition between rivals in this digital age is likely to grow even more fierce, and these days we have a lot more rivals than just Kodak. It's going to take a lot of managerial muscle to keep the company on a growth track.

Whether confronting a crisis or pushing ahead with structural reforms, one question keeps coming to mind: What makes a robust enterprise, a really strong company? The twenty-first century has been and will continue to be a time of dramatic transformation. The global balance in business is constantly changing. The market never stops moving. Economic fundamentals are impacted by the availability of financial capital. The exchange rate and raw materials costs fluctuate dynamically. And it is all fast—very, very fast. How to come to terms with these changes and navigate these troubled waters are the biggest issues facing twenty-first-century management.

For an enterprise to emerge a winner in times like these it has to be quick to adapt. More than that, it has to look ahead and anticipate the future. A *good* enterprise can do this. But the *best* enterprise is a company that creates change on its own. By offering a new mechanism, a new product, a new idea, it creates new value and changes the world. And it produces such products one after another.

Fujifilm must, as a manufacturer working with that goal in mind, continue its R&D to establish a foundation for future growth and constantly create products that reflect the values of the times. Examples of these efforts would be our American clinical trials under way on drugs for Alzheimer's disease, and Fujifilm's clinical trials for anticancer drugs that began in May 2013.

I see Fujifilm ten years from now as a company with its own original technology, a company that creates unique products at the cutting edge and offers them to the world. And I like to imagine—but I'm sure I'll still be surprised sometimes—what those products might be.

The Difference between Kodak and Fujifilm

Fujifilm's net sales for the fiscal year ending in March 2013 amounted to ¥2.2147 trillion.* I wish these numbers were higher, but they show that we confronted the critical loss of our core market and successfully embarked on a new course of growth.

The previous year, our industry experienced an event of epic significance. In January 2012, Kodak filed for bankruptcy under Chapter 11 of the U.S. Bankruptcy Code. This was the same giant that, when I joined Fujifilm, had more than ten times our sales.

How did a leading company—once the undisputed leader in the photo field—with Kodak's out-

---

* In 2013, ¥2.21 trillion = $21.4 billion.

standing technology, finances, branding, and marketing, sink so low? Most who watch the world marketplace believe that that American companies adapt much more quickly to change than their Japanese counterparts. But in the case of photographic film, this conventional wisdom got turned on its head.

"What made the difference between Kodak and Fujifilm?" I've been asked this question innumerable times by the *Economist*, *Wall Street Journal*, Reuters, Bloomberg, the *Frankfurter Allgemeine Zeitung*, *China Business Journal*, *Chosun Ilbo*, and others from all over the world.

First of all, one thing that might have contributed to Kodak's downfall was that it was the premier company in photographic film for so long. This, I believe, made it slow to adapt and diversify. Fujifilm, which was always the challenger, always pursuing Kodak's colossal shadow, needed to try its hand at diversification.

Of course, Kodak had also seen the coming of the digital age with the same sense of foreboding. For example, it too had tried to diversify into the development of pharmaceuticals, very much as Fujifilm had.

But Kodak was hesitant to modify its core business. From the outside, it appeared that Kodak deep down just really didn't want to. And though both companies did diversify, there was a big difference in the depth and breadth of our execution.

Kodak knew which way the winds of the digital age were blowing, but it sold off the pharmaceutical business and others it had acquired as insurance against the future, and retreated to photographic film, where quick profits could be made. This would later prove to be a fatal mistake.

Second, Kodak and Fujifilm confronted digitalization differently. Digital products weren't as profitable as film, but Fujifilm read the signs and immediately undertook their development in greater breadth and depth.

So, as contraction of photographic film accelerated from 2000 onward, Fujifilm assumed a leading position in the expanding digital camera market and realized pretty impressive sales. This revenue made it possible not only to cover losses incurred by the contracting film market, but also to accumulate the technology and know-how needed to ride out the waves of the digital storm.

A third factor was Fujifilm's digital minilab. Our compact digital printer/processor was placed in photo stores around the world to print out images from digital cameras on photographic paper. This allowed the minilab businesses to continue providing printing services in the digital age.

Fourth, Kodak withdrew early on from developing and manufacturing its own digital cameras to rely on OEM suppliers instead. Not having its own technology such as sensor and image processing put Kodak at a considerable disadvantage when the digital race began in earnest.

Certainly, letting a company whose bread-

and-butter business is making photographic film get involved with developing digital cameras sounds like a conflict of interest. The profits of one would inevitably cannibalize the profits of the other. So how to resolve this problem? How did Fujifilm make the leap?

The answer is simple. If we had not done it, another company eventually would. There was basically no choice. That we saw the answer and Kodak didn't obviously made a real difference in the resolve of the two companies.

Fifth, Kodak later declared that it was a digital company, or wanted to be one, and seemed to be evolving in that direction. But here again Fujifilm was different. Even though we had entered the digital age, Fujifilm had not the slightest intention of transforming itself entirely into a digital company.

By concentrating solely on digital, by engaging only in digital businesses like photography and imaging, the most we could hope for in sales would be several hundred billion yen. Differentiation in the digital world is difficult, and sooner or later price competition would heat up, meaning that even this smaller amount of revenue couldn't be relied on. Digital products alone are not enough to sustain a company worth several trillion yen.

Yet, for an industry colossus like Kodak—our mortal rival over so many years—to go bankrupt is a cause for genuine regret. I have no doubt that our competition made us—both of us—better over the years.

SIDEBAR

Disaster Reconfirms the Cultural Value of Photography

The Great East Japan Earthquake of March 11, 2011, wrought enormous destruction through the powerful waves of the subsequent tsunami. Fujifilm's main manufacturing factories escaped damage, but Fujifilm-related companies located in the hardest-hit areas did not, and we made every effort to bring them back on line. Immediately, on the evening of the disaster, our on-site sales reps began helping clients to get back on their feet.

The victims, in the midst of the devastation, frantically searched for the photographs that marked the important moments in the lives of their families. They could work and make money again. They could rebuild their homes. But the photos of their families could never be taken again; they were unique and irreplaceable. The devastation brought this truth home and made us realize once again that photographs can be, literally, priceless.

Fujifilm received many inquiries at that time asking how photos covered in mud and

Photo Rescue Project.

damaged by seawater could be saved. So we immediately began to study how to clean mud-covered photos, and eventually found that the best method was to wash them in lukewarm water.

Fujifilm then launched the Photo Rescue Project. Our team members visited some eighty devastated areas on a rotating basis, teaching local volunteers how to clean photo prints and providing them with a "photo rescue kit" that contained items needed for cleaning and drying photos, such as rubber gloves and pocket albums. This project was a substantial mission for Fujifilm out in the field, but the number of photos that needed rescuing grew so large that it proved impossible for people on a local level to handle them all.

Fujifilm then decided that the photos received from victims would be sent to our Kanagawa factory, cleaned and restored, and returned to them in the devastated areas.

Counting Fujifilm employees and alumni, their families, and others, a total of 1,500 people cleaned some 170,000 photos at the factory. Include the photos cleaned by volunteers around the country and the number rises to several million. Never before had I felt such a strong sense of mission—that the photos our products had made possible should be cleaned and restored by our own hands.

Another thing I learned from the disaster was that most of the photos rescued by the Self-Defense Forces and others were not in the form of digitalized CDs or flash memory, but were actual photographic prints. The pictures with the greatest likelihood of high-quality restoration were not those printed by home inkjet printers, where the ink was almost certain to run when exposed to water, but the ones that had been processed and printed at a photo store as silver halide photos with a surface coating—the way it was done when Fujifilm began.

Through the Photo Rescue Project we reconfirmed our belief in the invaluable role of photography in human life. From here on it will remain our everlasting mission to preserve the legacy of photographic culture, and pass on to later generations the wonders of photography.

PART TWO

Managing for Victory

MANAGING IN TIMES OF CRISIS

READ, PLAN, COMMUNICATE, ACT

Corporate management and national governance need improved leadership. The absence of a leader or a lack of leadership is not just a political or economic problem, but an issue that plagues every aspect of Japanese life. How is a leader to be defined? What is required of a leader? What capabilities should a leader have?

Consensus Leaders Are Useless

I've often been asked during interviews about the transformation at Fujifilm, "Didn't your drive for reforms cause friction and opposition within the company?"

My answer is, simply, no. Faced with a life-and-death crisis, there were no employees against reform (that I knew of, anyway), and even if there were, it wouldn't have made me hesitate to do what needed to be done. Anyone who worries about everyone else's opinion when the survival of an enterprise demands decisive action can't be called a leader.

In a sense, business is like a war, a fight to the death. In such a situation, what general on Earth would find it wise to ask the opinion of individual soldiers departing on a mission? And what soldier would question his general's orders when the enemy is just over the hill?

A company is an aggregate of organizations with special functions. The decision-making process of each organization is carried out through meetings of senior management and by the board of directors. Accordingly, the person who verifies decisions and makes the final choice as to the company's future direction is the commanding officer, the leader.

If a business leader decides to act, he does so with a full sense of responsibility and concentrates solely on those actions that will lead the company to victory. Even if there is some resistance, he resolutely does what needs to be done. That is the role of a leader, especially in times of crisis.

"Didn't you discuss things with company employees?" interviewers have also asked me. Of course, many business plans originated from sources close to the front line, and I did listen to and discuss matters with individual divisions. But when we were facing the most significant and perilous crisis in the history of the company, there was no way that I, as president, was going to become a consensus leader and make decisions by a majority vote. Again, it sounds harsh, but you can't govern by committee, and you don't have time to make consensus on a foundering ship.

It's the same on a battlefield where the fighting is fast and furious and you have to decide what orders to give next. No army battling it out on the front lines decides things by taking a vote. The commander assesses the situation and makes a quick decision: advance on the left; on the right; withdraw; attack; go for it! It is neither the time nor the place for a leisurely democratic discussion.

When things are quiet, it may be fine to make decisions in a friendly manner and by majority rule; but in a crisis, while everyone is discussing whether we should go this way or maybe that way, if you hesitate you give up the opportunity to emerge a winner rather than a loser. Someone has to lead the way, to make the tough choices. That is the role of the leader. In fact, it is the essence of leadership.

But leadership comes with great responsibility. A mistake could lead to the ruin of the company, and that is the leader's burden as well. If you were a samurai in the past, you would have had to commit ritual suicide. Today that is not a realistic option, but you still must take the fall. Now, as then, leaders are not allowed to fail.

Four Steps for Managers in Times of Crisis

So, what is the function of management? It's argued in business schools and debated by economists, but it is really quite straightforward. In brief, it is to provide products and services with social value, to create sales and make a profit, and, through investing in the future, to ensure the continued survival of the company. You have to consider diverse practicalities.

During a crisis, when you are caught up in a battle that will determine the outcome of the war, the steps a manager must take can be reduced to four. These four are always required, regardless of who the leader is or the situation that's in front of you.

Reading
At the top of the priority list is "reading," in the same sense an American football team analyzes the opposition before the game. You have to read the situation

you are in. Right now, at this very moment, what is it you are confronting? What exactly is happening? With limited time and limited information, the leader has to gain an accurate understanding of the current state of affairs.

Another kind of reading concerns the future, what is going to happen next. As the situation changes minute by minute, the leader has to make the best possible estimate of what the future holds, based on his or her understanding of what is happening at the moment.

Planning

Once you have a good read on the present and future, use that as a basis to consider the direction the company should take and what it should be doing, formulating this as strategy. The plan is your script for executing the strategy.

Communicating

The focal point of a company confronting a crisis may be the strong determination of its top manager, but the manager can't overcome the crisis all by himself. Your resolution must be communicated to the farthest reaches of the organization; your sense of crisis must be shared by all; each and every employee has to be fully aware of his own position and purpose.

This is where communication plays an important role. Your message must cover the same things that would be mentioned in a battlefield briefing:

What is happening now? What might happen next? Where are we headed? What has to be done?

Acting

A CEO is not an academic or a critic. It is not enough for him simply to say, "This is the present situation; this is what the future holds; and therefore this is what we should do." Once you make a decision, you have to carry it out. Without action and execution, all your reading and planning are worthless. You firmly take the lead and see that your decision is implemented and your orders carried out.

* * *

Whether the crisis can be overcome and whether reforms will meet with success depends on how faithfully and precisely the leader carries these four steps out.

Let's take a closer look at them.

Understanding the Present Situation with Limited Information

What does "reading" the present state of affairs really mean? In a time of crisis, the leader's first move is to understand precisely what is happening. To do that, he must carefully examine all incoming information.

In the throes of a crisis, however, getting accurate information isn't easy. It does not appear as a neat little package on your doorstep. It is invariably fragmentary and incomplete. Sometimes information is concealed or falsified, both from without and within. There are times when new information comes to light after a decision has already been made and there is no turning back. That has been my experience more than once.

So a leader cannot be satisfied with just reading the information available on the surface; he must also dive deep, read the fragments, get a picture of the whole and grasp its essential nature: "This is wrong; that looks true; this appears about right."

Anyone who thinks this is too much trouble isn't qualified to be a leader.

Of course, when others get their hands on accurate information they will easily see what's happening, too. By then the value of that information has been considerably diminished. A true leader must do his own reading early in the game.

How about the information's accuracy? Do you see pieces missing? What does your experience tell you it all means? Paying painstaking attention to these details, the leader can gain an accurate picture of what's really going on. Commitment and devotion to this task is a true test of leadership ability.

As for myself, I don't claim to have my own personal information network. I seek out general information from newspapers, magazines, and books. Information specifically related to our company

I obtain from in-house reports. For example, the monthly reports submitted by the various businesses and divisions contain market data, current problems, news about competitors, sales performance, and more. It's a lot to absorb, but Fujifilm is a big company. I've gotten accustomed to it over the years, so now just a glance tells me what is important or crucial. The unimportant I can skip.

But more critically, after years of reading such material I've gotten so I can read into it what is *not* there. Authors, without being aware of it, let their underlying motives and intentions seep into their reports. So even when something is stated explicitly, you can't necessarily take it at face value. In fact, the meaning may be quite the opposite. Sometimes an author may be trying to cover up an error that was made before.

You can recognize this kind of writing by its sudden leaps in logic, with the problematic passage standing out from the rest. There may be a lack of confidence apparent in the writing style—or a definite sense that the author is trying to hide something. While sniffing out these dubious passages, the leader has to discover the true meaning of the report and gain an understanding of the situation. If there is something I wonder about, I immediately place a call and ask for clarification.

At Fujifilm, I have asked that important passages be underlined within material meant for the eyes of upper management—but in fact, this underlining isn't all that useful to me. Instead of reading what the writer or his manager thinks is important, you should focus on the parts that *you* sense are important. So even if a sentence is underlined, I may ignore it if I think it's just not vital.

I have also asked that divisional reports be summarized on a single sheet of standard letter-size paper. This helps the author produce a precise (and just as importantly, concise) report, and it helps the reader quickly grasp the principal points. It is all about management efficiency.

Reading the Flow of Events and Predicting the Future

The second kind of reading is the kind that predicts the future. Once you have a proper grasp of the present state of affairs, you take that as your starting point and ask, "What is going to happen next? What is unfolding here?" You read the flow of events.

The accuracy of their predictive reading is a big factor in differentiating leaders from another. In World War II, France capitulated to Nazi Germany, but Prime Minister Winston Churchill refused to surrender. Even while London was being bombed every day during the Blitz, Churchill continued to call for the English people to resist to the bitter end. He was able to do this because he had an accurate reading of future developments. He knew that if England kept fighting, the United States would eventually have to enter the war. He also figured that no matter how powerful Nazi Germany might be, it would then be outmatched. And just as Churchill predicted, the United States joined the war, and the Allied Nations, including England, went on to victory.

In the case of Fujifilm and its structural reforms, I conducted my predictive readings many times to determine how the digital age would make its appearance and how long the effects of the global economic downturn would last.

For these kinds of insights, you need forecasting based on precise numbers. Management, in many respects, is all about numbers and reading what they mean. How much is the market going to shrink? What is going to happen to profitability? With good numbers in hand, you can conduct simulations repeatedly as the situation evolves, and get a very good read on the future of the company.

I'm not saying that in our case these readings were always correct, but I believe they were generally on target. I did my best to raise their accuracy to improve the quality of our forecasts.

When mining newspapers and magazines for information, a manager has to study not only what is on the page—you have to think about what's going to happen next and do your own predictions. Let's say a newspaper article reports that "Company A has announced such-and-such." I consider the possibilities of this announcement and what is behind it. Why did Company A make this announcement? Why now? What are the real factors behind the announcement?

Maybe a few days later—or sometimes much later—I get my answer in another article or other bit of intelligence. Sometimes my reading was right on target; other times, it was way off. When I've miscalculated, I reflect on what caused me to be wrong. What did I miss? I turn the problem over in my mind.

If you do this sort of thing long enough, you begin to gain an instinctive understanding of the reasons behind certain events. "If A happens, B will also happen. If you fail to do C, D will inevitably crop up." You begin to see the universal law of cause and effect. If you can build up examples of this kind within your own experience, you can quickly and accurately predict future developments from the slightest sign, even if times and conditions are different, and even though the players may have changed.

Called upon to make readings in a limited timeframe, top managers must continually improve their skills. These days I regularly go through five newspapers. When I first became president and was pushed to the limit, I would sometimes let a week's worth of reading pile up. But I never failed to catch up. Even now, I don't read just what's on the surface but ask myself, "Why did this happen? Why are they doing that?" I question what lies behind the information presented, and I reflect on what is going to happen next.

Though I've been at it for decades, I keep practicing the art of "reading," and not just for my own edification. I also keep working on my ability to read the big picture, the entire course of events. You cannot let up on your information analysis if you want to compete on the world stage.

Applying Universal Laws Outside Your Area of Expertise

I wrote above about the universal law of cause and effect and the importance of internalizing your understanding of it. I cannot emphasize this enough. Your ability to do so is crucial not only for reading the future, but also for making judgments outside your area of expertise.

For example, I studied economics and am not particularly schooled in technology. But does that mean I know nothing about technology? No. As the top manager at Fujifilm I am the one who makes the go/no-go decision on any business venture. To do that, I have to have ask and answer such questions as: What technology will the future require? What kind of products can such technology produce? Do we have the technology today that we need to ensure our growth in the future?

How can an economics graduate answer questions like this? Engineers have asked me, "How is it that, even with something technically challenging, you get to the heart of the matter so quickly?"

It's true that I don't always fully grasp technical discussions or specialist jargon, but it's also true that everything behaves according to some universal principle or law. This is the case not just for the social sciences, but also for technology and hard science. In every endeavor, the law of cause and effect—"If A occurs, B will follow"—is applicable.

So when you see something new, you ask, "Does this fit with our prevailing notion of cause and effect, or is it contrary?" If A happens, is it going to be followed by B? And if the answer turns out to be C, why?

If you are always thinking along these lines, whenever you come across some innovative technology or unusual phenomenon, you can check it against the law of cause and effect and reach a fresh understanding of what lies at the heart of the matter. This way you won't be taken by surprise. You will know true from false, the possible from the impossible, and understand what should be done next.

Three Ways of Misreading the Present and Future

Of course, you can still miscalculate, and if you do you have to use it as a learning experience. Looking back over various experiences and incidents from my past, I see three principal ways that you are likely to make mistakes, either in how you understand the present or what you predict for the future.

The first kind of misreading arises from not facing reality directly. An accurate reading requires a dispassionate eye. Reality has to be accepted for what it is. There can be no looking aside. It was sad but clear from the figures that Fujifilm's photographic film market was most likely going to shrink to almost nothing. It took courage to face this reality, even if it was supported by statistics; but if we had not faced it, all our future decisions would have been compromised.

The demand for the product we had devoted our lives to was rapidly disappearing. This meant we would all be out of work, which we understandably found hard to accept. Fortunately, division heads and upper management arrived at an objective, dispassionate view of the situation, and their conclusion was eventually shared and solutions implemented on a companywide basis.

The second kind of misreading occurs when you receive one-sided information. When the information you have comes from only one source, or you possess only one type of data, the chances of a misreading are much greater. It is best to gather different kinds of information from a variety of sources—and the more reliable the better.

For example, when you are thinking of starting a new business, you should get input from multiple and diverse sources pertinent to your plans: technology, manufacturing, sales and marketing, logistics and more. Of course, if your sources have a slanted view, your judgment based on that information will end up being slanted as well. You have to cast the net wide for high-quality intelligence.

I have already mentioned the episode where I decided to invest in and build up the polarizer protective film business as the market for liquid crystal displays expanded. At the time, I paid particular attention to gathering information from a variety of sources, since we didn't know yet whether liquid crystal or plasma screens would have the larger market.

As a manufacturer of components, we had our own view of the matter, but that wasn't enough. We visited the companies in charge of product assembly and sought out their opinions on both liquid crystal and plasma. After combining this with other information we had gathered, I came to the conclusion that liquid crystal would win.

Failure to gather all the information you need before you need it inherently skews your knowledge base. You have to really work at reporting, mining your contacts, and driving the discussions, so that later you don't end up saying, "I should have asked about that, or I wasn't aware of that."

The third kind of misreading occurs when you are influenced by a preconception—a bias or a predilection. You think, "This just can't be happening." Or, "A market this big can't possibly be shrinking." If you're inclined to rely on your prejudgments, you can't possibly view things objectively. You may think, "This seems to be right, but I really don't want to do it." Or, "Since this was proposed by so-and-so, we'd better pass." This type of thinking invariably leads to misreading your options—and to potentially disastrous missteps in your business.

Deciding Priorities and Drawing up Realistic Plans

The second step of a leader is planning. It's important to understand present conditions and peer into the future, but if that's all you do, you're no better than a TV analyst or pundit. Once you've made your readings, you have to decide your priorities and draw up a plan of action.

The leader has the responsibility for deciding the direction of the company, to steer it to the right or the left. In what fields should we create new businesses? What kinds of new products should we develop? How can we win the current battle? Top management decides for the company as a whole; the department head for the department; the section chief for the section. Without respecting this hierarchy, the organization will not function.

Determining priorities is vital here. What absolutely must be done? What is of the utmost importance? No matter how difficult something is, if it is the top priority it has to be done. Japanese managers tend to postpone decisions when some other vested interest is involved and opposition is strong. A leader who can't make decisions is someone with a low sense of priority. This is the worst type of leader—in fact, not really a leader at all.

Your decision about what to do next becomes abundantly clear if you have done your homework in understanding the present and predicting the future. If sales have gone down and don't respond to your attempts to revive them, for example, you need to reduce the size of the organization and its facilities. Prioritizing your actions and reactions enables you to execute your plan.

Dynamism and Speed

When Fujifilm was faced with the severe contraction of its core business, its top priority was to create another source of income that would make up for the lost revenue in the film market. We drew up practical plans for what we needed to do, and at the same time we began to downsize.

In a situation like this you have to be constantly aware of dynamism and speed—the timing, the pace, the scale. A good plan that should have gone well will fail if you are slow in taking action or your timing is off. The same is true if your response is too small. When a patient can only be saved if a diseased area is removed by surgery, operate quickly and do it all at once. If you cut in bits and pieces, the repeated surgeries will weaken the patient.

The same is true in war. Using your forces piecemeal in follow-up operations will only lay them open to being picked off one at a time. To accomplish a goal, to win a battle, you can't get the timing wrong, and when you act, you must act decisively.

The situation at Fujifilm during the reforms wasn't easy. Shoring up existing businesses, starting new ventures, and, at the same time, restructuring—all this was going to be expensive, and we would have to make unfortunate employment decisions as well.

Downsizing was something none of us wanted to do. I could understand why people wanted to put it off or do it gradually, but in the end, a leader has to act deliberately and with conviction. A piecemeal approach would only result in the company growing weaker as time dragged on.

That is why I made the final decision after carefully considering the scale, the extent and the timing. If I had been slow to decide and deal with the dwindling film business, the company might have been swallowed up in the ensuing recession.

The huge investment we made in the production line for polarizer protective film is another good example. The proposal that came to me called for the creation of a single production line. As I listened to the argument I sensed that the investment should be bigger than just one line. So I asked, "If we invest in other lines in the future, when would that be?"

"Maybe a year from now," came the answer.

Well, if that was the case, it ought to be done all at once, I decided. That way we could both shorten the delivery time for equipment and cut costs. In the end, I decided to put in four lines at once—that was my read of the market—and it turned out to be very timely as the demand for LCD televisions exploded.

Your first premise must be that your idea of what you want to do is correct. After that, the most important thing for you as a leader is to be correct about the speed, timing, and scale.

The Need for Muscle Intelligence

Why do so many managers get speed and timing wrong? Why do they dither when faced with a crucial decision? I have seen many managers act this way, even though they were very intelligent people.

Once you have made a decision, don't hesitate to follow through. This has nothing to do with brain power, but requires a certain type of visceral instinct, intuition, and some strength. I call this "muscle intelligence."

For example, if you are caught in a fire, which direction and how fast should you run to escape the flames? You won't find the answer in a textbook and it has nothing to do with academic ability. The difference between the people who escape to safety and those who don't is not based on intelligence; it is a difference of instinct and intuition.

Whether a person has this ability or not is a big factor in who succeeds at the top. I don't mean to sound immodest, but I have been blessed to some extent with this type of instinct. When faced with a crucial choice, I often get an inspiration about which is right, and I'm usually not wrong. I would point to Fujifilm today as proof.

This ability is not something you learn through studying. It is gradually developed and refined during childhood when you are swimming in the ocean, or diving, or mountain climbing, or playing samurai, or quarreling with other kids. If you had to develop your "muscle intelligence" only after reaching adulthood, you would probably be able to do it through your everyday work, by becoming highly attuned to your tasks and especially alert to speed and timing. It's never too late to try.

Even When You Hesitate, Make It a Success

When I became CEO, I wanted to be like an all-powerful force with infallible judgment. You may think that's the height of conceit, but it was more like a wistful hope. In reality, I thought it would be useful to have a dominant measure of control, superior intellectual capacity, and broad competence. Or so I felt at the time. I knew, too, that perfection is a futile pursuit.

As Fujifilm's CEO I thought that if I made a hundred decisions, I would have to get a hundred decisions right, and I strove body and soul to do that. I thought that's what was needed in order to save the company. But working like this every day really gets you down. Whether I was awake or asleep, the company was always in my thoughts, ruminating compulsively, and some nights I couldn't sleep at all. There was a mountain of decisions waiting to be made, and all had unyielding deadlines.

The point is that becoming the top manager was definitely not a walk in the park, no matter how ready I felt for the job. There were times when I agonized until the last minute but still couldn't come up with a decision. There was just no clear difference between the two choices in front of me.

What can you do in a situation like that?

If you have some time to spare, put the issue aside for a while. Sometimes your subconscious mind will find the answer. You might wake up out of a sound sleep and think, "*Yes!!*" Or you may be thinking about something else altogether when a clue to solving your problem suddenly pops into your mind. The subconscious is always at work; even when you're not actively thinking about your problem, your brain is still processing it without pause in the background.

But when you don't have any time to spare, what do you do then? I think about it this way: Maybe the reason you can't make up your mind is that there is no big difference between the two choices. Maybe both are correct.

After racking your brain over the problem, what you should definitely *not* do is put off the decision.

Doing that won't get you or your enterprise anywhere at all. When the time to decide has come, you have to decide, no matter how hesitant you feel.

And once your decision is made, you make sure it leads to success. Whichever way you choose, you give it everything you've got to make it work out in the end. That's all there is to it. Even when the outcome is unclear, you make the decision, confident that it's correct, and take the lead in achieving success—and then you actually do achieve success. That's what it means to be a leader.

Keeping Refreshed and Invigorated

Leaders are called upon endlessly to make decisions, and most of those decisions have to be the right ones. In a crisis, the importance and frequency of the decisions multiply, exhausting both body and spirit. And letting fatigue build up will have a negative impact on your decision-making.

I said there were three factors leading to misjudgment, but I should add that not being in the best physical and mental condition can lead to hesitation and mistakes, too. I have failed to make decisions when they should have been made because I wasn't feeling in top condition. No matter how blessed you are with good instincts and intuition, they won't function properly if you're not in good health. And no matter how physically fit you are—

no matter how strong your desire to get the job done—your judgment will be impaired if you are mentally exhausted.

A leader has a duty to keep refreshed and invigorated. In a time of crisis the word "refreshed" may sound a little lackadaisical, but in fact, the deeper the crisis, the more the leader needs to keep in good physical shape, remain vigorous, and maintain the conditions conducive to making the right decisions.

Even in the midst of the most grueling daily schedule, I always kept in mind the need for being refreshed. Reading a favorite mystery before going to sleep, enjoying a drink, and playing golf relaxed me and provided a change of pace, clearing my mind for making decisions.

Most important of all is a stable, loving family. There is no better way of refreshing oneself than relaxing at home.

Without Communication from the Top, the Organization Won't Budge

After reading the present and future, formulating plans for what is to be done, and making the final decision to go ahead, the leader must then communicate it all in a clear-cut message to employees. If the message is received and understood, the organization will begin to move in the new direction of its own accord, its own internal momentum.

The reforms we called the "Second Foundation," described in Part I of this book, were made possible at Fujifilm by the capabilities of every employee. And disseminating information from top management in the right way and at the right time was crucial to raising the morale and sense of mission of front-line staff.

In addition to sending a message to all employees four times a year via the in-house newsletter, I explained the company's vision and goals at fiscal-year meetings at the main office and at principal factories and research centers, as well as during my New Year's speech. In these talks, I repeatedly explained the present circumstances of the company, the direction we were headed, and what we needed to do—or more correctly, what must absolutely be done—toward achieving those goals.

If you conduct proper briefings and indicate clear strategy and goals based on a logical analysis and assessment of the way things are, everyone can move toward those objectives without too much fret. No matter how tough the going, people will persevere if they know what they need to do to get the job accomplished.

The leader really has to give some serious thought to how he is going to communicate with employees—hold a townhall on the factory floor, talk to them one-to-one, write something for the in-house newsletter, give a speech. Depending on the situation, the time and the place, he has to choose the most effective means available.

When Fujifilm undertook our second series of reforms following the financial crisis of fall 2008, we held small meetings with front-line staff to explain policy matters. It was entirely different then from the situation when the photographic film business was impacted by digitalization. After 2008, not just one, even if it was the biggest, but every division in the company was affected.

To get the message across, I thought it important to explain it in the simplest, most direct way possible. What was the scale of its impact on the global economy? What part of the future can we predict? What cannot be predicted? My point was that to accommodate ourselves to the future, we would have to build a robust company that could withstand a shrinking economy and market.

In groups of ten to twenty on-site staff, managers created an atmosphere of interactive communication and spoke directly to every employee. I decided to use this same style when conveying my message both directly and indirectly to each person. By doing so I was able to quickly and clearly relay the intentions of management and the significance of the reforms. That the post-crisis changes could be achieved so rapidly means that we got the communications right.

Leading Is More Important Than Thinking about How to Lead

Once the present and future readings have been made correctly, the goals and plans formulated, and the direction of the company communicated to the employees, the only thing left is—action! If action doesn't follow your plans and your words, then all your efforts have been for nothing.

If someone else had been president during Fujifilm's reforms, that person may have made the same readings I made, conceived the same plans, and communicated with the employees in much the same way. However, in carrying out the reforms, I hope that I was more thorough and resolute than others could ever have been.

When it comes to action, the leader must stand at the head of the troops, take the initiative, and set an example. In the words of a famous Japanese admiral: "Do it and show it, say it and tell it, have them try it, then praise them and they will do it by themselves." When you come right down to it, the first thing for a leader to do is to show everyone how the job is done. Without that example, neither the organization nor the employees will move.

A famous story about setting examples involves Marshal Ney, one of Napoleon's leading generals. The Napoleonic army had lost the war in Russia and was withdrawing, with Ney in charge of the rear guard. The enemy was massed directly in front of them, and no one wanted to lead the chilling charge into their midst, but someone had to do it. Then Marshal Ney calmly said, "Okay, let's go," and spearheaded the assault. As luck would have it, he escaped unscathed. The art of war is strange in that way.

When leading the action, you may have second thoughts about your troops: "Are they really going to follow me?" But if the leader has properly conveyed his message, the organization will come together and move toward the agreed-upon goals.

A new section chief came to me once for advice, "How should I go about leading people? I don't know what to do." You shouldn't be wasting your time thinking about that, I told him. Make your decision and push ahead, no matter what anyone says. That's what it means to be a leader. If you do that, people will follow. If you have enough time to worry about how to lead, you should instead apply that time to thinking about reading the situation and making mistake-free plans for action.

If you are thinking, "Maybe no one's going to follow my lead…," or "What if I'm wrong?" it shows on your face and causes those around you to be uneasy. Just make the smartest, most intelligent decision possible and set an example, and the employees will naturally follow.

SIDEBAR

Number Two Uses a Bamboo Sword, Number One Uses Steel

I often say that the top manager fights with a steel sword, whereas the No. 2 manager—likewise the managers below him—uses bamboo.

When the fighting is with steel swords, to lose means to die. There is no learning from mistakes. For the top manager to lose means that the company has lost—at least for that round. That marks the end for the top manager, but it also means that the company is seriously wounded. That's why mistakes are simply unacceptable. And that's what produces the desperate need to win.

As president, this is the attitude you have to take. You can't get away with saying, "Oh, everyone makes mistakes." It bears repeating that in the old days, the samurai paid for their mistakes by committing ritual suicide, or *seppuku*. A top manager has to adopt a similar do-or-die attitude.

The burden of responsibility for the organization's top manager and the person just below is far different. No. 2 has responsibilities befitting his position, and if he makes a mistake, there is always the top manager above him. But if the top manager commits an error, the whole organization will go into a spin. The difference between the two is immense. That's why the top manager is always working under tremendous stress, fighting with steel and not bamboo. And that's why losing is out of the question.

A BATTLE THAT CANNOT BE LOST

THE WAY OF THE WORLD AND THE STRENGTH TO COME OUT ON TOP

No surrender!

In business as in life, my mindset has always been one of refusing to give up. This same resolve is a cornerstone of the Japanese psyche. Not just the reforms at Fujifilm, but both life and business are for me a kind of battle. And for me, losing a battle is not an option. To win, however, requires preparations, and adopting a strategy.

All Life Is a Battle to Be Won or Lost

A fundamental principle of society is that everything in life is a struggle, a continuing battle. Here as I see them are the main contests we face:

The battle against rivals and adversaries.
The battle against the current of the times.
The battle against destiny.
The battle against hardship.
The battle against custom and tradition.
The battle against personal weakness.

Every battle ends in either a win or a loss. Of course no one wants to be on the losing side, and that is why we have to train ourselves physically and mentally, and psych ourselves up for the fight. Following the actual battle is either victory or defeat, advance or retreat. This is how things are, and we must face facts.

I bring this up now because there is a tendency in today's world to avoid conflict. Conflict produces winners and losers—this is a cruel fact of life. But just because you are afraid of becoming a loser and being hurt, or becoming a winner and hurting someone else, doesn't mean you can avoid conflict forever.

The Japanese these days seem to fear friction and discord. They tend toward the school of thought that says, "Let's all just get along and play nice together."

I am sorry to report that even at Fujifilm when I say, "Get in there and fight," some look at me blankly and ask, "Fight?"

Under the banner of our democracy, we believe we can discuss things amicably as we go along, without anyone ever getting hurt. This is utterly impossible. Why? Because life is a battle, a struggle for survival. Even those who have until now managed to avoid conflict will one day find themselves doing battle with a ruthless newcomer to the scene, a veritable tiger. Such fights are inevitable, because the world has become an arena of ferocious competition. On the day of battle, the lackluster camp won't have a chance. And there are today more and more such people, born in lukewarm, conflict-averse waters, who have never experienced the pain that comes from losing.

Young people are very responsive to suggestion and will carry out orders to the letter. But is that enough in today's fiercely competitive world? The true competitive spirit, the notion that "I'm going to win, no matter what," is gradually fading away. You can't win if you don't believe you can win.

We have to develop a more competitive, more challenging spirit. We have to build rock-solid strength. If we don't, global competition will simply blow us away. Everywhere there are battles to fight, and everyone, whoever they are, must accept this fact of life, and must then strive day in and day out never to lose a single battle. This holds true for business as it does for everyday life.

Postwar Japan Teaches Me the Wretchedness of Losing

To me the business reforms undertaken at Fujifilm were a battle that I could not concede. I faced the most critical crisis since Fujifilm's founding, the loss of its core business, with the conviction "There's no way I'm losing this fight!"

However, I didn't come to believe that "life is a battle I refuse to lose" only after I became president of Fujifilm. My experiences in postwar Japan brought me to this conclusion as a very young man.

I was born in Mukden City (now Shenyang), Manchuria (northeast China), in 1939. I was five when World War II came to an end in August of 1945. The unrelenting conflict between Japan and the Allied Nations had reached a conclusion and Japan had lost. The scenes I saw then are still embedded in my mind.

Russian soldiers, the last to enter the Pacific war, would shoot their automatic rifles into the air and then break into houses and make off with watches and precious metals. Plundering, looting, pillaging, and ransacking were rampant; anarchy ruled the day. And the Japanese, the losers in the war, were particularly easy targets.

My father worked in Mukden, where he had a Western-style residence that doubled as a business establishment. He had no choice but to give it up and try somehow to get back to Japan. In the chaotic days that followed, we hid out in underground shelters and abandoned houses, and tried somehow to survive.

Then one day my father handed me a short Japanese sword. He didn't say anything, but I instantly understood what he meant: "If something happens to me, it's up to you to protect your mother and little sister." Even without a word from my father, protective feelings for my mother and sister rushed into my chest. At the same time, though but a child, I hated the thought that Japan had lost the war.

Getting back to Japan turned out to be an extremely arduous undertaking. After spending a day and night being jostled along in an open freight car, we reached an internment camp where we had to wait a month for a ship. Food there consisted of nothing but red sorghum porridge and a soup made of potato tendrils. Later I learned that a good many children had died in that camp.

It was in September of the following year that we finally reached home in Sasebo Port, Nagasaki. Manchuria had been bad, but Japan was no better, with most of its principal cities now in ashes, razed to the ground. Our only possessions were the clothes on ours backs and our rucksacks.

As young as I was, I couldn't help but feel in my very bones the misery that comes to a defeated nation. The suffering, anguish, and tragedy that accompany defeat were driven home in the most excruciating manner possible. It was this primal experience, I believe, that played a big role in forming my personal desire to acquire an unrelenting might.

Building a Bedrock of Strength to Escape Defeat

Refusal to lose became the fundamental mindset in my life as I grew up in Japan after the war.

I was able to get into the University of Tokyo thanks to that way of thinking. At the time I took the university entrance exam, it was commonly said that you could pass if you slept only four hours a night and studied the rest—but if you slept five, you would fail. This was called the "4 pass, 5 fail" rule. I even heard stories of students studying with a large tatami needle stuck into their thigh to stay awake. That's how fierce the competition was back then. Passing the university entrance exam represented a big battle at this stage of my life. That I was able to fight my way through it was due solely to my belief that failure was simply not an option.

But I didn't think that just getting into a good school was all I needed to do. Throughout my life, the most important goal has always been building a foundation for personal strength. Even as a child I thought about what things I could do to make myself genuinely strong.

What would that foundation consist of? On an individual level, a strong mind, a strong heart and a strong body. Later, on a company level, the best workers, an outstanding corporate culture, top-notch technology, solid finances and super sales. Without developing these areas, no matter what knowledge or techniques I possessed, I couldn't really be strong—in life or in business.

To build this foundation, I read a lot of books in high school, aside from my regular studies. Even while studying for university exams, I remained engrossed in books. I read widely in world literature. I knew instinctively that exam prep alone wouldn't provide the foundation that I needed.

I also knew that building a strong body was important. From my youngest days I was always engaged in some kind of physical activity. From playing samurai to swimming, track and field, baseball, tennis, or judo, I was constantly doing something to build up my body.

The year I entered the University of Tokyo, a team was just starting up to compete in an American-style college football league in Japan. Thinking, "Just what am I waiting for?" I joined the team to maximize that all-important physical foundation of strength.

Football requires at least five things: fighting spirit, physical strength, speed, strategy, and teamwork. All these elements can also be applied to daily life and to business. Even after graduation and finding a job, and after my competitive arena had changed to business and management, I have relied on what I learned from American football as a basis for my battles in life. The impact of football on me was that intense; it forged a physical and mental foundation for genuine strength.

Never one to be distracted by immediate tasks or the pursuit of quick fixes, I have always kept the following questions in mind as I worked:

> *What can I do to grow as a person?*
>
> *What should I study to build a stronger foundation for genuine strength?*
>
> *What can I do to produce even greater results in my work?*

The fundamental strengths acquired during your life become your own heritage, a broad collection of personal assets. The same thing can be said of a company's history and culture. Developing and accumulating fundamental strength and genuine power is of utmost importance. Once you have done that, no matter what happens, no matter what you endeavor to do, you can expect to achieve good results.

The Whole Body Theory of Business

What decides how much a person actually accomplishes in his or her work? After considerable experience in a working environment, I have my own theory on the subject: The performance of a person on every level is the sum total of his or her "human power."

Whether your business performance is good or bad, whether you are a success or failure, whether you win or lose, all depends on the totality of your human capabilities. I call this the "Whole Body Theory of Business." In effect, these capabilities are the very foundation of any businessperson.

Let's take them up one at a time.

Eyes, Ears, Nose, Skin, the Sixth Sense
To produce good results before your rival does, you must have the ability to acquire intelligence faster and more accurately. You must sharpen your senses of sight, hearing, smell, touch and taste. You must quickly grasp the information circulating in the world and, even if it's fragmentary, see into its essential nature. At times you must transcend the traditional five senses and exercise intuition, the sixth sense.

Head (Brain)
The head refers to your capacity to take the information acquired by the five senses and analyze it, reformat it, grasp its essence, discover problems, and formulate strategy and tactics. Any qualitative differences here have a huge impact on battles being waged in the business world.

Chest (Heart)
Here the chest means the heart. Do you have a fundamental interest in others, are you capable of sym-

pathy and empathy, do you have an accepting heart? Are you capable of love? Do you have a strong sense of mission, of corporate ownership? Without these attributes, you will not be willingly welcomed into the group. Without them, no one will follow your lead. Having a heart makes a big difference in winning or losing.

Stomach (Guts)

Do you have courage? Do you have guts? (Or "intestinal fortitude," we sometimes say.) There is the idiom "to have the stomach to do so-and-so." That is what we are referring to here—having the courage to make a bold decision, to brace up, to step up to the plate, to carry through against all odds. This all depends on "guts."

Legs and Hips

This refers to your ability to take action. You may have a good head and good heart, but they are useless without action. Even when you make a decision, always having someone else carry it out is a sign of weak legs and hips. The will to go to the front lines; indefatigable, immediate strength; grass-roots perseverance—these are all indications of strong legs and hips.

Arms and Hands

Here we're talking about technique and skill. And when there's a need to muscle your way through a dif-

Theory of the Total Business Person

Learn humbly from everything and follow an upward spiral of continued growth.

The job performance of a person is the sum total of his/her "human power."

Eyes, ears, nose, skin, sixth sense (1)
Grasp the truth.
Fathom the essence.

Head (mind) (2)
Make use of every opportunity and formulate winning tactics and strategy.

Chest (heart) (3)
Be fundamentally interested in others and capable of fellow feeling.

Stomach (guts) (4)
Have the courage to make a decision and put it into action. Carry through against all odds.

Face and posture (8)
Increase your capacity as a human being.
Develop character.

Mouth (7)
Say what you mean.
Communicate.

Arms and hands (6)
Polish techniques and skills.
When necessary, push forcefully through.

Legs and hips (5)
Go into action.
Pick up the pace and move quickly.

ficult situation, arms and hands refer to sheer power.

Mouth

This refers to your ability to express yourself, to precisely convey your thoughts. It symbolizes the ability to persuade people to your way of thinking, the ability to communicate.

Face and Posture

What is important is not whether you have a beautiful or handsome face, but whether your face conveys a firm sense of principle, mission, responsibility, and confidence. Posture refers to whether your backbone is straight and erect. Facial expression and posture mirror your inner life and are outward signs of whether you have improved upon yourself in the ways discussed in this book. An unpleasant facial expression and poor posture invariably create a bad impression.

* * *

Work performance and business success depend on the sum total of your human capabilities. If one element is missing, the totality will be reduced, results will not follow, and defeat will ensue. This is my Whole Body Theory of Business, and it is particularly important for leaders. After all, it is through their capabilities as total human beings that top leaders are able to engage each individual employee and lead the company as a whole.

Keep this in mind in every possible situation and continue to refine and improve your abilities. Areas of strength you make stronger, and areas of weakness you work hard to fix. This applies both to you as an individual and to you as a company leader. As you grow, you and the company can undertake larger and larger projects; and with what you learn from those projects, you and your company will achieve even greater growth. This kind of upward spiral is essential.

Without Gentleness and a Cause, "Winning" and "Strength" Are Meaningless

In whatever you do, you have to win, and to win you have to be strong. That is certainly important. But winning and strength have no meaning in themselves. A leader or a corporate warrior should be fighting and growing strong for some greater purpose.

Maybe it's for your country, for your organization, for your family, or for a larger cause; and if you are a businessman, for the company, for your sense of corporate mission. "Cause" here means something of greater significance, something of overarching value. To be strong in order to defend something of such all-encompassing worth is real strength, the strength that leads to a meaningful victory. Some people are strong only for themselves, but I can't find much meaning in that kind of strength, and I know for a fact that even when these people turn out to be winners, it doesn't last long.

In my Whole Body Theory of Business, what I consider particularly important is the chest or heart. I am a fan of foreign hard-boiled mysteries, and fifty years ago I read one by Raymond Chandler in which private detective Philip Marlowe has the famous line: "If I wasn't hard, I wouldn't be alive. If I couldn't ever be gentle, I wouldn't deserve to be alive." I am in total agreement with this.

Of course, it is not enough just to be gentle. A great many Japanese these days seem to possess this quality, but when it comes to personal strength, many are found wanting. For example, when something goes wrong, they don't confront the results unblinkingly and assign responsibility, but leave the whole matter enveloped in a cloud of smoke, with the parties involved consoling one another. This is the type of lukewarm tenderheartedness you may find in Japanese socicty today, but it can't really be so kind or tenderhearted. What is kindness without strength?

Winning through Understanding International Behavior and Manners

No one in business can ignore how work and management are conducted abroad. There are several international behavior patterns that Japanese business people must know to succeed outside of Japan.

These patterns may be more properly described as linguistic etiquette. It is a matter of responding and expressing oneself in the most appropriate manner. This is not a Japanese strength. Whether a response is positive or negative, it must be in keeping with the behavior or language of the listener. Unless you get this right, you are likely to be made fun of and dismissed as a lightweight.

Consider the matter of expressing your opinion. Japanese can be described in positive terms as gentlemanly, but they can conversely be perceived as too quiet. If a Japanese businessman says nothing, it's taken to mean that either he has no opinion or he is in total agreement.

When I was in Europe I often saw Japanese at meetings silently taking notes while the foreign business person was talking up a storm. So I say to employees stationed abroad, "After you give the other party ten minutes to express their opinion, take ten minutes to express your own." Japanese often think that if they express a contrary opinion they will make the other party angry, so they hold back. This is a common and unfortunate misunderstanding among my countrymen.

When I was stationed in Düsseldorf as president of Fujifilm Europe, I said what I thought without mincing words. I proposed the development of new products, had my say in pricing matters traditionally decided in Tokyo, prepared a new sales campaign, changed the corporate culture, and kept urging staff to become No. 1 in the industry and dominate the color film market.

Throughout Europe I ordered the distributors in each country to prepare a strategic plan for reaching the top within three years. Since, up until then, Japanese management had generally satisfied itself with issuing vague directives, the Europeans were somewhat taken aback. They seemed to think, "No Japanese has ever said anything like this to us before."

Some of our distributors didn't present a plan, and I had to push them hard. I warned those who still showed no intention of cooperating, "Are you going to do it or not? If not, be prepared for the worst."

If you're stationed abroad in this dog-eat-dog business world, being able to assert your opinion when crossing swords with non-Japanese colleagues is vital. You have to present your thoughts and views as clearly and unambiguously as possible. Anyone who just keeps quiet will be seen as lacking the courage to speak up, as incompetent, or as just plain unqualified and incapable of standing on his own two feet.

To a certain extent, Japanese can communicate among themselves without verbalizing their thoughts. They expect to be understood without actually speaking. But this won't work abroad when dealing with a non-Japanese person who was brought up in a completely different environment. If anything, you are more likely to create a bad impression.

This is why it is so important to express your own opinion—why it is crucial to make your thoughts known if you want to create a positive relationship. Outside Japan, if you don't stand up for your own point of view, you will be lost in the fray.

Of course, if you lack a logical basis for your argument, you won't be taken seriously. You earn respect only when you negotiate from a solid factual position and with proper logic. Do that, and you gain the respect of the other party and find them more willing to compromise. From this interplay emerge mutually creative ideas and real friendship.

And you also need to support your position with details. For example, just saying, "Do your best" is meaningless. You have to get all your ducks in a row—explaining precisely why and how to do something. At Fujifilm Europe I didn't just set the goal of becoming the No. 1 company in the industry, I also told people how to do it and gave them the tools by which it could be done. In an international setting, everything has to make sense in this way.

The first given is that you speak English—it remains the international language of business. After that comes expressing your opinion and making sense. Anything else is spurious—just meaningless words and behavior.

I often told the resident Japanese staff, "Japanese may not be tall, but they have big hearts and brains." With this and a fighting spirit, Japanese can compete successfully abroad.

Winning Intelligently, Honestly, and with Spirit

My father taught me the importance of fighting. If I came home after being picked on by an older boy, my father would give me a whipping and tell me to go back and finish the fight. From my youngest days this notion was beaten into me, so to speak—a man doesn't give up and he doesn't lose.

The whole world is a battle that you cannot lose. The lesson I learned as a child will remain with me for life. But don't misunderstand me. I am not saying that winning is the only thing that counts. That would make us the same as the animals. For example, I don't endorse the strong bullying the weak.

Some say that battle and competition strain human relations, creating a world where the strong tyrannize the weak. I disagree. In school sports, children compete for first or second place on their own merits and without being bullies. Bullying exists outside of such formal and respectful competition.

It is not battle and competition that is the cause of people trying to take advantage of each other, but a lack of moral education. Yet moral education is not that complicated. For example, my parents taught me these very simple but very important rules:

> *Don't do anything underhanded.*
> *Be fair.*
> *Be honest.*
> *Don't lie.*
> *If you lose, don't cry.*
> *Don't bully the weak.*
> *Don't cause inconvenience to others.*
> *Stand up straight.*

These rules should be sufficient, I think, to govern most moral behavior. On the Japanese island of Kyushu where I grew up, strictures against being cowardly were particularly strong, and that rule of behavior, along with the others from my parents, formed the moral foundation for my life.

These same rules can also be applied to business and management. But if you take the view that anything goes in business as long as you make a profit, you will eventually lose the support of others—your customers as well as your colleagues. Good management needs to be unfailingly based on what makes human beings human: wisdom and goodwill, ethics and aesthetics, and a sense of dignity.

When you win, you must win intelligently, honestly, and with spirit. You must fight fairly and make your opponent think, "Well done!" or "If we have to lose, I'm glad it's to this company!" In corporate enterprise, competition must be fueled by a special underlying sense of mission, such as contributing to society or working for the good of the nation.

When I use the word "win," it might be better interpreted as "overcome" or "prevail." "Winning" connotes that one party comes out on top and the other loses, a zero-sum game. But to "overcome" doesn't suggest that, since the word can also be used in the sense of "overcoming" one's own weaknesses. So, while it's important to win, it is also important to win intelligently, honestly, and with spirit, to "overcome" rather than just to "win."

SIDEBAR

Books I Have Read to Build a Foundation of Strength

A foundation of strength is a human being's most powerful tool. It is what allows us to face reality directly, to think, to feel, and to act. Our study of economics may be useful when it comes to making a living or doing our jobs, but if our own underlying foundation is weak, whatever knowledge we have cannot be used effectively. First, we must build a foundation of strength, and only then can we tackle practical matters. This sequence is extremely important.

How do you acquire this fundamental strength?

One way is to turn all the various events and incidents encountered in daily life and at the workplace into learning experiences.

Another way is to read good books, whether philosophy, history, or fiction. When training leaders, the British give priority to philosophy and history. Next would come familiarity with fiction and other forms of literature. What I have found most useful in my own life are philosophy and history. These subjects help the reader cultivate a sense of the past, build a broader perspective, and assess value systems—essential tools for making critical decisions.

Thinking that I would build up a foundation of strength, I read collections of world literature from beginning to end, from Tolstoy and Dostoyevsky to Stendhal, Balzac, Romain Rolland, Ogai Mori, Soseki Natsume, Ryunosuke Akutagawa, and others. I had finished reading nearly all of them by high school.

Here are some of the books I particularly recommend.

Since both business and life are themselves a kind of war, if you intend to become a top leader, you cannot ignore history, particularly the history of war. I was forty when I first read *The Second World War* by England's great wartime leader, Winston Churchill. It is the account of his trials and tribulations as prime minister during the war, and how he led Great Britain to victory. This work is largely responsible for his receiving the Nobel Prize in Literature. I learned a great deal from this book and have read it two or three times.

The lessons it contains are many. What is occurring at the present moment? How will things develop? Churchill's analysis of the state of affairs at that time is remarkably tough-minded. And once he decided what the English should do next, he was resolute in converting that decision into immediate action.

Churchill famously said that man must keep an eye on reality because reality is keeping an eye on man. How a man confronts reality reveals his fundamental strength. It is a mirror image of his real capability and performance. Looking at stark reality without blinking, and having the backbone to then challenge that reality—this is the true spirit of a leader.

History is a profound study of the nature of mankind and also teaches us much about the evil side of human beings. Many unsavory characters appear in the annals of time.

The Chinese histories *Record of the Three Kingdoms* and *The Eighteen Outline History*, which I read after joining Fujifilm, depict the dark side of life—the traitors, the liars, and other contemptible types.

Another book that left a lasting impression was Eiji Yoshikawa's *Miyamoto Musashi*, which I read just after entering high school. I was deeply moved by Musashi's indomitable spirit as he relinquished a hedonistic life, exercised strict control over his behavior, and sought to perfect himself as a human being by pursuing the art of swordsmanship. Up to that point I had led a fairly carefree life, but then I came to think, "This is what a man should be," and "I have to exercise self-control and perfect myself." It was after reading this book that I became seriously and fully engaged in whatever task I undertook.

There is one more book that changed my life—the novel *Jean-Christophe* by Romain Rolland, which takes Beethoven as its model. The story depicts a man who overcomes many setbacks in becoming a successful musician. After reading it, I thought, "I refuse to lead an idle life with no meaning."

When I entered the university, I began to devour philosophy books one after another. The philosopher most akin to my way of thinking was Friedrich Nietzsche. I read most of his works, including *Thus Spoke Zarathustra, Beyond Good and Evil*, and *Ecce Homo*.

Live a life unimpeded by strictures and be free! Live strongly, honestly, and nobly! By throwing off the shackles of old customs, entrenched authority, and established regimes, what is truly wonderful in the human being will come to the surface. Nietzsche called the person who was capable of this the *Übermensch*. His philosophy fit perfectly with my thinking at that time. I too, I thought, must be free and live strongly, honestly, and nobly.

THOSE WHO PUT THE COMPANY FIRST ARE THOSE WHO TRULY GROW

HOW TO BE SUCCESSFUL AT WORK AND CONTINUE TO GROW

Working in the same place at approximately the same age, and at same type of job, some people will grow, achieve great things, and become leaders, while others will not. What is it that sets them apart?

There may be differences in ability or intelligence, but more important is the difference in the attitude they have toward their work and their feelings toward the organization they work for. This holds true for whatever industry, and for whatever country.

The person who grows through his work, what kind of person is he or she? The person who produces results, how does he or she do it?

The Company Is Not a Classroom

I recently heard that one of the criteria university students use to choose a company to work for is whether that company will help them grow as employees. Sorry to bring you the news, but it doesn't work that way. The company is not a kindergarten, a high school, or even a graduate school; it does not hold daily classes for your edification or enjoyment. Educating employees is often important to a company, but in no way its principal function. And even if employees take advantage of the learning opportunities a company provides, that doesn't necessarily mean that they will benefit from them and grow—which, of course, is their own problem.

Fundamentally, any growth achieved by human beings is self-growth. The company declares the objectives, points the way, and gives detailed instructions. And then employees must exercise ingenuity in carrying out their assignments, accomplishing their tasks, and, through that process, learning on their own what needs to be learned. Only by engaging in his or her own work with a strong commitment to develop and progress through self-growth will the employee ever become an improved person.

Looking at the process in another way, the best employees are indeed the ones who rise to the top on their own power, but management can't simply wait for that to happen. This is the purpose of on-the-job training and why employees are shifted around to find the right person for the right job. Management is strategically trying to create circumstances for employees to rise up from below.

Everyone has unique strengths and weaknesses; each person has different abilities. Some are good on offense, some on defense. Some are sales-oriented; some are meant for administration. Finding just the right position for each employee to play is vital.

But most important of all is that any education provided by the company, even if not the classroom kind, be strict and exacting. You might even call it a Spartan education. Just speaking softly to employees and being mealy-mouthed won't help anyone grow at all. Recently the theory has been that it's important for managers to praise their employees. And that's true. If you give a subordinate a task to do and he performs well, then go ahead and praise him: "You did a great job!" But if he really misses the mark, then it's just as important that you be critical: "You did a poor job here."

I still watch football on television now and then, and generally speaking, it seems to me that the teams with kind-hearted coaches are the weakest. The stronger teams are those with the tougher coaches. This stands to reason since football is a kind of martial art, and I think of management in much the same way.

I am convinced from my experience that it was under the stricter managers that I grew the most.

That's not to say that I didn't grow under managers who were appreciative of what I was doing, but it was the tough ones who brought out the best in me.

Being strict means you actually care about your subordinates' future. Only when you are dealing straight from the shoulder can you deliver a stern reprimand or take a tough stance. I'd go so far as to say that lenient managers who are afraid to scold their employees are irresponsible. Because they don't seriously believe in their subordinates' ability to grow and improve, they are incapable of giving a scolding when it is most needed. The employee, the manager and the company all suffer.

Learn from Whatever Comes Your Way

One prerequisite of your personal growth and development is your ability to learn from whatever comes your way. Errors, mistakes and painful experiences are ideal occasions for learning. Your future growth depends on how much you are able to absorb from these trials. I have benefited from many such occasions.

Near the end of the 1980s, I became director of Fujifilm's Recording Media Division, which handled sales of videotapes and floppy disks. The field was thick with famous brand names, and as soon as I arrived on the job, I was greeted by a fierce price-cutting war. A two-hour tape selling for ¥5,000 dropped ninety percent in price to just a few hundred yen. We had failed to use higher quality to differentiate our product from the others.

This turned out to be a bitter experience for me, but at the same time I learned a lot from it. The first thing was that nothing is more meaningless than price-cutting that has gone too far. If the race to see who can sell at the cheapest price continues unabated, every company in the competition will be hurt and end up without the resources to invest in future and better products. Continued progress in quality improvement will stop, and the consumer will be the loser.

To avoid this and survive into the future, we needed to develop a product that could rise above the others and sell because of its superior performance. So Fujifilm created a high-quality tape using a double coating—one magnetic layer for audio and another for video.

This lesson proved its worth more than ten years later when Fujifilm took the lead in developing a digital camera and saw its market share grow. Rival companies would sooner or later come out with their own cameras, and it was easy to predict that price-cutting would ensue. I knew from experience that excessive price-cutting had to be avoided. So we relied on our corporate strengths, creating appealing products, differentiated by quality and function, and developing a marketing strategy that didn't involve price-cutting competition. Still, the digital camera market remains an area of fierce price wars. As I write this, Fujifilm is pushing its X-Series, cutting-edge digital cameras with a price befitting their superior value.

The lesson is that when something happens, evaluate the consequences. If things have gone well, you try to figure out why they went well. If they haven't gone well, you reflect on the reasons for failure.

Those who learn from every event or experience, and keep what they have learned in their minds and hearts, will without a doubt grow as people and as workers. They will put what they have learned to use in whatever they take up next. This will produce good results in their work, and once again they will learn from the experience.

The result here is a positive growth spiral that should be the goal of all businesspeople hoping for success.

Work with a Sense of Ownership

In my twenties, I was not what you would call a model employee. I frequently butted heads with my boss. Then one day I happened to be assigned to a particularly open-minded manager. When I spoke up in my usual manner, saying exactly what I thought, he told me, "That's the good thing about you. Keep it up," and he assigned me a big job to handle.

I felt deeply moved, and I swore to do my very best for this man. I had to repay him in some way for believing in me. A sense of mission welled up in me, and at the same time I began to seriously think not about myself but about the company. What would make Fujifilm strong? What could my department do toward that end? And what should I personally be doing? I had never thought so intently about these things before.

From that experience I became convinced that there is one additional criterion for personal growth: a strong feeling for the welfare of the company, a strong sense of ownership.

The person who has these qualities thinks not only of himself and his own welfare, but is someone who works twice as hard because he is working for the company, the organization, the team, for his colleagues and the people above and below him. It takes a great deal of hard work to get a large number of people—whether a team or a corporation— to function smoothly. It is much more difficult than just working for yourself. It requires the maximum amount of effort, but it is this very effort that forces a person to grow.

I went on to become a manager, a senior manager, a general manager, and president of Fujifilm Europe, and I then returned to Japan to become president of Fujifilm. During all those years I was constantly thinking of what I could do for the company, what would be best for Fujifilm. My top priority was not myself but always the enterprise. Was I working with a sense of ownership? Was I contributing to the company? These questions were never far from my mind. This sense of mission spurred my motivation. And that in turn provided the engine for growth.

Strong companies owe their strength to their many employees who think first of the company, their department or their unit. These people don't think of themselves, but always of the company and their particular role, and that's why they strive so hard to produce the best results day in and day out. And it is these people who, transforming their sense of mission into motivation, raise themselves to higher levels of personal growth and performance.

When Chairman Onishi appointed me president, I never asked the reason why. But one day I heard from another party his assessment of me: "No matter what happens, he is always thinking of the company. He always has the company foremost in mind and acts accordingly." This, I think, is one of the reasons I was appointed president.

As for myself, when I judge employees, I always consider how much of a "company person" they are. You may have ability, but if you are the selfish type who thinks only of yourself, you won't go far in my organization. When faced with the question of "for the company?" or "for myself?" those who choose the company will invariably be the ones to grow into their jobs.

The notions of working devotedly for your company, or fighting loyally for your country, may not be very popular in Japan these days, but I still believe it is only natural to love your company, your country, your family, and your friends. It is only natural that you should be willing to work like the devil for your loved ones. It is this sense of noble duty that provides the impetus for people to grow.

Take Whatever You Do Seriously and See It Through

One thing is of supreme importance in life: sincerity and honesty in your dealings with yourself and others. Whether it's your work, your company or your family, do this wholeheartedly, not holding back, giving it your all. This kind of sincerity, this type of single-mindedness, is essential.

You may think this is just common sense and hardly worth mentioning, but the fact is, whether a person is productive in his work or not, whether he grows as a person or not, doesn't depend solely on his innate ability. In most cases, it depends on whether he is sincerely engaged in his work and giving it everything he's got.

Some time ago, when visiting a branch office, I had the opportunity to go out for a drink with some of the local employees. One of them, a young man in his thirties, said to me: "I don't know how to balance work and personal life. How much energy should I put into company work?" My answer was: "Devote all your time and energy to the company for six months. Then you will know." After a while I heard from this young man again: "Now I think I understand," he said.

No matter how much you think about them, there are some things that you just can't see clearly. But if you engage in them sincerely and single-mindedly, they assume a clearer form. That form is the truth. If such commitment is beyond you, then you won't learn anything and you will never grow.

Such sincerity and perseverance are the Japanese people's greatest strength. A sense of ownership also plays a part, but what's really important is that there is something of great significance that transcends individual interest, coupled with a spiritual and physical stamina that allows you to emerge victorious in the end, no matter how tough the going gets. Once a person has really decided to do something, he or she doesn't mind burning a little midnight oil.

This particular attitude toward work is something Westerners seem to admire about the Japanese. In fact, Western workers have often told me: "What I find most amazing about Japanese people is that they firmly believe that hard work will produce success!"

After the Meiji Restoration in 1868, Japan strove to rival the Western powers as an independent country, and then as time passed we actually became their equal as a world power. After the conclusion of World War II, Japan rebuilt itself and became an economic power. The moving force behind this achievement, as Westerners themselves recognized, was the Japanese devotion to hard work.

Westerners for their part are good at devising efficient means of reducing their workload, of producing results without burning much midnight oil. Up against relentless Japanese hard work, they contrive ways of winning through strategy, through method and technique. Japanese are often quite the opposite, not being as focused on strategic thinking. This is an area that obviously needs improvement.

Before Relying on Others, Ask Yourself What You Have Done

When I was young and working in the Graphic Arts Systems Division, some products were not quality-competitive and just didn't work all that well. I had the opportunity at a gathering to speak with President Kusuo Hirata, the fourth-generation president at Fujifilm.

In the foolishness of youth I asked him directly: "The product quality is not good. Can't you get the factory and the research people going on this?" He answered, "Yes, I'll tell them you said so." But then he added something even more important: "But tell me, what have you been doing yourself on that score?"

I immediately realized what he meant. Before asking someone else to do something, you had to do something yourself. What could I do to get the factory and the research division to improve quality? The only way I could think of was to involve them more in the process, so I persuaded some engineers to accompany me on my rounds visiting customers. I hoped that if people from production and development heard, face-to-face, what our clients had to say, that would do the trick.

It wasn't easy, but somehow I managed to convince the engineers to come with me. Today it's not unusual for engineers to accompany sales people on their rounds, but at the time it was hardly ever done. First I took them to a well-meaning, but critical and sharp-tongued client. When I asked him for his thoughts, he immediately brought out a competitive product and said, "Look at the difference here. There is no comparison." We were the also-rans.

Hearing with their own ears what our clients had to say lit a fire under these engineers. Bluntly facing such severe criticism, they were bound and determined to develop something even better than our competitor's. Needless to say, product quality rose.

After this episode, no matter what the problem, I took it personally, not as something unconnected to myself or to be relegated to others. I asked myself what I could personally do to solve it. I learned that it's not enough to say upper management is at fault, or your colleagues or subordinates. You have to take action yourself.

In whatever you do, don't take the easy way out by relying on others. Instead, find what you can do and pursue it with everything you've got. Your attitude here is yet another factor in whether you grow as a person.

Without Changing Reality, There Is No Progress

Let me tell you another story from my younger days. I was in my late twenties when Fujifilm began to sell high-resolution photographic plates for electronics manufacturing. Kodak had already preceded us into the field, and it was hard to find anyone who would use our product.

When I asked prospective clients what we could do to get them to choose our plates instead of Kodak's, their response only added to our problems. They wanted us to resolve various difficult issues that Kodak hadn't overcome in its own plates yet. It was impossible for us to fix all these problems at once, and consequently they declined to even give our product a chance.

At my wits' end, I finally got some advice from the head of my department. "The point is, you have to get them to try it, if only once," he said. That's when it struck me. Just trying to convince them by talking wasn't going to work. I had to change reality to make any progress. So once more I visited our customers. And this time I vowed I'd convince them.

"In any case, try it—just once," I said.

"Just for a little while is fine, but give it a try."

"If you try it out, our engineers may notice something and be able to improve on it," I offered.

I'm not sure whether my impassioned appeals themselves turned the trick, but from then on they began to use our plates. From this I learned that just talking doesn't move the situation forward. When you have hit a wall, you have to change a reality or two. It might lead to a breakthrough, and you may score a win.

Why Some Senior Managers Don't Grow

It has long been said that noncommissioned officers in the Japanese army are the best of the best. But it is also said that once they become commissioned officers, they go to the dogs. They can't see the big picture, they can't develop strategy, and when a large-scale problem occurs, they just aren't up to it. In the face of important decisions that require courage and resolve, many tremble at the thought.

Unfortunately, this propensity lingers on in Japanese business—and maybe worldwide business, too. Someone who up to now has been working like the devil becomes a senior manager and suddenly stops growing.

Why does this happen? There are a number of possible reasons, one of which is physical condition and health. The person in question has been working relentlessly for years and has simply run out of gas.

Another reason is that the person is satisfied with the status quo, with the position he has reached. We hear a lot these days about young people becoming lethargic, indifferent, and apathetic, but the same thing can be said of long-tenured senior managers.

A third reason is that maybe the manager didn't train himself earlier for the day when he would reach the upper ranks. Training means preparing yourself physically and mentally. It means reading history and philosophy, thinking about the big picture, and broadening your perspectives. A leader without the big picture in mind or without a sense of values finds it difficult to make decisions, and the same applies to leaders lacking courage.

Managers who find it hard to decide something, who content themselves with simple stewardship of a department or unit, and who have lost the will to do anything else, certainly did not get the message when they were more junior. Today they're afraid of dealing with big problems, putting their own fears above the company's success.

In war, your ordinary soldier may be outstanding, but if his officers are not, you cannot hope to win. The same is true in a corporate setting; here, too, junior employees must be trained so they can eventually exercise leadership in upper management. One way of doing this is by sending promising managers abroad to experience the tough environment in the outside world. In my case, serving as president of Fujifilm Europe proved enormously educational.

Short of posting someone abroad, for a person to serve as the president of a subsidiary can also be enormously rewarding. Being responsible for everything from sales performance to business control and management brings out the best in a person and makes him or her a bigger, better individual. In the end, the load shouldered by someone at the very top compared with someone slightly lower in the hierarchy is entirely different in both quantity and quality.

SIDEBAR

Not Plan-Do-Check-Action, But See-Think-Plan-Do

In order to bring about the "Second Foundation" of Fujifilm, the medium-term management plan dubbed VISION 75 proposed three policies: building new growth strategies, implementing companywide structural reforms, and enhancing consolidated management. These policies needed employees who were highly motivated and who possessed enhanced skills. To bring individual employees to the top of their game, we devised an original developmental approach called the Fujifilm Way.

The core of the Fujifilm Way was the Whole

Body Theory of Business mentioned earlier. Two points gained from broad conversations with on-site managers were added: the aspects of Fujifilm that should be preserved and the bold reforms that should be initiated. This was all put together in a booklet.

In the future, Fujifilm expected to move into new businesses and come into competition with companies from different industries, requiring a different approach to business. We had already begun to enter areas where the standard rules of thumb and long-established knowledge no longer applied. Moreover, the speed of change was dramatic and interrelated factors were increasingly complex. The future would demand productive results even when the information needed for decision-making was unavailable. The Fujifilm Way was created to deal with this situation.

The Fujifilm Way booklet established STPD (See-Think-Plan-Do) as the basic management cycle. This is in contrast to the widely known and accepted PDCA (plan-do-check-action). Fujifilm wanted to improve on PDCA by giving more emphasis to the preliminary stages,. So we inserted See and Think before Plan and Do, as follows:

See—Think—Plan—Do

1. Collect information.
2. Analyze the information and bring issues to light.
3. Set goals and aims.
4. Draw up a winning scenario.
5. Draft a practical plan of action.
6. Carry through to the end.
7. Review and sum up (gather feedback for future reference).

See: Prioritizing "what" and "why" without immediately rushing to "how."

Think: Seeing into the heart of the matter without immediately conceptualizing.

Plan: Building a solid, stable plan.

Do: Executing on the action, resolutely and boldly.

In an age of dramatic change, the STPD management cycle is far better than PDCA for producing results when challenging new, unknown areas of business.

THE WAY FORWARD IN A GLOBAL AGE

CORPORATE AND NATIONAL STRENGTHS

Things have changed considerably since the 2012 introduction by Prime Minister Shinzo Abe of his economic reforms package—known as "Abenomics"—but for a while after the global financial crises of 2008, Japan as a whole was enveloped in a paralyzing fear of the future. Even the mass media went to extremes, depicting Japan as having suffered an irretrievable breakdown.

But was Japan really done for? No such thing.

Naturally, it is not all smooth sailing, but the country continues to boast superior technological prowess, skilled workers, and great potential for growth. But first we need to understand the present situation; if we do not, it's likely we'll take the wrong path forward.

Japan's Manufacturing Sector Is Not Losing Ground

Twenty years have passed since the bursting of Japan's economic bubble, and during most of that time the news was rife with stories about how the Japanese had lost confidence in themselves. After the events of fall 2008, the crisis facing the manufacturing industry—the engine behind our postwar growth—similarly became a frequent focus of the media.

Cutting-edge products that Japan in the past would have been the first to develop have instead been developed abroad, and in some cases have dominated world markets. Countries thought to be technologically behind Japan have started to catch up. "Japan is finished," the media say, but is that really the case?

It's true that Japanese products have failed to become the world standard in some areas. Here and there, Japan has lost market share. In the long term, however—and with a broader perspective—the affected areas are very limited in number. Yet the media go on as if the entire Japanese manufacturing industry has gone to ruin, which is highly doubtful.

In the current long, drawn-out depression, newly emerging countries and advanced nations alike have begun buying what is cheap, even if the quality is mediocre. To make things worse, due to the extraordinarily strong yen, foreign goods are thirty or forty percent cheaper than Japan's, enabling foreign companies to use this advantage to make remarkable advances. This is hardly a case of Japanese manufacturing suddenly growing feeble.

With a few exceptions, I am not a bit worried about Japanese manufacturers. Japanese manufacturers—not just large corporations, but also small and medium-sized companies—still have the developmental capability and technological strength to produce high cost-performance products that will be attractive to emerging nations. There is still a great deal of Japanese technology that can compete on an international level. Japan's manufacturers are full of untapped potential.

Yet some problems remain in connecting this technology with product development and positioning. Small and medium-sized companies have difficulty selling their products on the global market. But these issues are far from intractable. Price competition and leveraging technology into products that consumers and businesses want and whose value they will pay for can both be resolved with a more forward-looking approach.

In addition to its traditionally strong hard goods such as cars and electronic products, Japan has also many exportable systems for the industrial, infrastructure, and service sectors. Japanese convenience stores have gained a worldwide reputation for their high product quality and well organized store management at more than thousands of stores. And there are also softer, more intangible products that can be exported. For example, there are great possibilities in areas like waterworks management systems and other semipublic industries.

When we look at specific areas, it becomes clear that Japan and Japanese industry still have a great deal of potential and a bright future. We shouldn't be fooled by the media's constant drumbeat that Japan is finished.

The Slow Economy Due to a Strong Yen

Contrary to what the media have claimed, Japanese manufacturing has not died. It did experience some very hard times—the principal culprit being the strong yen. In the 1980s, Japan's manufacturing industry had become so strong that the major world powers sought to rein it in by handicapping Japan through adjustment in the currency exchange rates. This intervention in foreign exchange marked the beginning of Japan's economic slowdown. It was called the Plaza Accord, for the New York City hotel where it was signed.

Japan was indeed strong. In 1987, the opportunity arose for the best Japanese companies to meet with the best American companies and exchange information and opinions. I was chosen to represent Fujifilm. Over a period of two weeks, I visited

Minnesota, New York, New Jersey, Florida, Missouri, and Georgia, and held discussions with major American companies.

The aim of this tour was to discover where the respective strengths of American and Japanese companies lay, but what most forcefully struck me at the time was the thought that Japan had already won the battle—and perhaps the war.

During the high-growth postwar period, Japanese companies had hired a great many university graduates, and consequently it was possible for all their employees—factory managers, store managers, department heads, section chiefs, and ordinary workers—to process the same high level of information. Japan had, in other words, established a superlative system for high-level decision-making, which was carried out only after on-site managers had shared information with their subordinates and made adjustments to deal with the various problems facing the company.

In the United States at the time, one leader at the center moved the entire company. A factory manager might write something on a piece of paper, hand it to a subordinate and simply say, "This is for your team to do." Without any exchange of opinion, everything was decided by the leader, and the organization moved accordingly. It was strictly top-down, with a huge chasm dividing management and worker. Such a system in itself is not necessarily bad. With an outstanding person at the top, it can prove quite effective. The problem is that outstanding leaders are not always available.

The situation in the U.S. has changed considerably since then, and not all American companies were as I describe them, but the top-centric structure was true of the many companies I did have the opportunity to observe. I was completely convinced at the time of Japan's overwhelming strength continuing into the future.

It was this very strength that gave rise to the Plaza Accord. The Accord was signed in 1985, a little earlier than the period I am speaking of, but even then I'm sure the United States was cognizant of Japan's power. That is precisely why it championed the policy of handicapping Japan through a strong yen.

Japanese industry took head on the staggering force of the strong yen. The exchange rate had been ¥360 until the yen grew undervalued in 1971 under the fixed-exchange-rate system that was abandoned in 1973. From a rate of ¥240 to the dollar in the 1980s, the yen eventually appreciated to ¥200, ¥150, and then ¥140, and by April 1995 it had fallen below ¥80 to a dollar. In just ten years the value of the yen had tripled. Given this upheaval, Japanese industry could scarcely compete on the world stage.

Japan and Japanese companies were now effectively contained by the strong yen. The Plaza Accord was the beginning of the twenty, and now nearly thirty, "lost" years that were to follow.

The effects of the strong yen were not confined

to the manufacturing sector alone. Many companies found they were now unable to compete in exports and moved their factories overseas. Partly in line with its policy of placing manufacturing near its markets, as well as revenue pressure from the exchange rate, Fujifilm, too, has now moved over half of its facilities abroad.

It is an economic fact that if offshoring is taken too far, domestic production will suffer. The domestic economy will shrink and in turn, investment will slacken and employment will suffer. What's more, there will be a decline in personal income, and tax revenue will drop. A slowdown in domestic economic activity should therefore come as no surprise when factories and jobs move overseas.

Having been forced to move abroad to survive, Japanese companies were eventually confronted with another difficult decision concerning their domestic operations—changing the terms of employment. This meant reducing the number of full-time staff and increasing the number of non-traditional workers. The assembly industry in particular tried to absorb its losses from the appreciation of the yen by instituting a two-tiered system of full-time and temporary workers. The postwar concept of lifetime employment in Japan saw meaningful changes.

The Japanese economy continued to struggle under the shadow of the strong yen and the effects brought about by the increase in temporary workers. The final blow came with the merciless appreciation of the yen following the financial collapse in fall 2008. The yen jumped by thirty or forty percent against the dollar and the other major currencies, and the Japanese manufacturing sector found itself plunged into an even deeper abyss.

Why did Japan accept the Plaza Accord in the first place? The Accord wrought terrible destruction on the Japanese economy. If Japanese leadership in the 1980s had been more politically astute and able to find some middle ground in the negotiations—even if not rejecting the Accord in its entirety—it might have reached an agreement more to Japan's advantage.

Just recently, under the leadership of Prime Minister Abe, the process of correcting the strong yen is finally making some headway. This is good news for Japan's manufacturing sector and will produce the conditions for fair international competition. In terms of purchasing power parity, an equitable exchange rate, which I consider to be absolutely essential, is said to be about ¥105 to ¥110.

When talking about the problems plaguing Japan, I have focused on the economy, including foreign exchange, because most of the difficulties facing nations throughout the world have their origin in poverty and economic issues. Both for the nations of the world and for their peoples, it is the economy that is at the heart of the matter.

Separating TPP and Agricultural Issues

Japan a few years ago made the decision to participate in talks on the free-trade agreement known as the Trans-Pacific Partnership (TPP), which have been ongoing, but the country still seems to be of two minds. There are a few facts to be aware of on this divisive issue.

The first thing to know is that most customs duties in Japan have already been removed. Agriculture and certain regulated industries remain. The businesses that foreigners are restricted from conducting are virtually nonexistent.

If Japan had chosen not to participate in the talks, exports to any of the member nations (Australia, Brunei, Canada, Chile, Japan, Malaysia, Mexico, New Zealand, Peru, Singapore, the U.S., and Vietnam) would incur a duty—a great handicap for the exporting nation. The duty could be four percent or five percent, and Japanese industries would have to raise their prices accordingly and compete on disadvantageous terms.

Japan is an industrial nation. Its population is not only shrinking but also growing older, resulting in a smaller domestic market, so the only way for Japan to remain economically viable is through exports. Joining TPP and working toward abolishing duties on products exported from Japan represents the only path forward.

However, the issue of agriculture, including the

challenge of self-sufficiency, is far from a trifling matter, as well. In Japan's case, the issue revolves around the customs duties levied on imported rice, meat, and dairy products. These markets will have to be opened, and Japanese farmers compensated by some other means.

The overriding concern here is the future of Japanese agriculture, even if its principal products remain under the protection of high tariffs. The average age of full-time farmers in Japan is now said to be sixty-seven. What will the situation be ten years from now? The present state of affairs cannot be maintained.

Regardless of whether Japan joins the TPP, Japanese agriculture will have to be revamped. It will have to become larger in scale and more competitive, perhaps through incorporation as in the United States. It cannot survive as it is.

Japan is blessed with brilliant sunshine and abundant rainfall, and a great deal of its land is suitable to farming. The problem is that 400,000 hectares, almost a million acres, now lie fallow.

China and other countries are buying up land at an increasing rate in the Mekong Delta and elsewhere in Asia and in South America. They are taking preventive measures against a future food shortage. To counter this, the United States and wealthy nations in the Middle East are also apparently moving in the same direction.

With this in mind, Japan must rethink its stance on agriculture in more strategic terms. It must take proactive steps and change the way the business of agriculture is conducted. Japanese agriculture today is an ¥8 trillion business, but it receives nearly ¥5 trillion in financial support.* We need incisive agricultural policies, including those that address the subject of fallow fields. However you look at it, and TPP notwithstanding, the reform of Japanese agriculture is an issue that cannot be disregarded.

Issues for Japan: The High Cost of Corporate SG&A

The strong yen is the root of most of Japan's economic problems, and without correcting it there is little hope for a quick recovery by the manufacturing sector. While Japanese manufacturers are blessed with top-shelf technology and highly skilled workers, and remain full of potential, they do need to face up to a few other problems before any future growth can take place.

One of these is the high ratio of selling, general, and administrative expenses (SG&A) to net sales.

This is a matter of considerable concern to Fujifilm. We once carried out a comparative study with a well-known American chemical company. This chemical company had a comparable structure of six business units and similar net sales of ¥2.5 trillion.

---

* In 2014, ¥8 trillion = $76.9 billion.

But compared to the American company's operating profits of over ¥500 billion, Fujifilm's at the time were only about ¥150 billion. What could possibly account for the difference? Well, at the heart of the problem was the appreciation of the yen. Calculated at the pre-recession exchange rate and raw material prices, that ¥150 billion in profit would have been about ¥225 billion, thus accounting for ¥75 billion of the difference.

Beyond that, there was also a divergence in R&D as a percentage of sales. Fujifilm was investing far more than the American chemical company, about three or four percent of sales. As mentioned earlier, American and Japanese companies adopt a different stance toward investment in R&D. American companies are forced to keep a close eye on short-term stock prices and are therefore very conscious of making a quick profit. R&D often suffers. Japanese companies are more concerned with the long view.

Investment is meaningless if it doesn't produce results, but if the seeds aren't planted, there will be no future fruit to pick. When R&D investment is cut, profit will show an immediate increase, as I noted before, but it is hard to justify cutting investment in the future as the right thing to do. A company cannot be considered top-flight simply because it is highly profitable.

Still, the low operating profit of Japanese manufacturers cannot be entirely explained by their higher investment in R&D. In our case, the difference between ¥150 billion and ¥500 billion in profit was hardly due to a three-and-a-half to four-percent R&D investment.

What causes the difference then? It comes down to SG&A. In most Japanese companies, SG&A accounts for twenty-five to thirty percent of net sales. In Fujifilm's case, it is twenty-six percent. In the case of the American chemical company that we benchmarked, it was no more than twenty percent. There was an astounding six percent difference between the two companies in this area.

If the four percent for R&D investment is added to the SG&A figure, the difference now becomes ten percent, explaining the gap in our operating profits. The combination of R&D investment and SG&A costs produces an overwhelming difference in the rate of return for American vs. Japanese companies.

So why should Japanese SG&A costs be so high? To put it simply, there are too many white-collar workers. The number of administrative employees is excessive; many companies have more back-office staff than production workers. Divisions such as Corporate Planning have become inflated, and some companies have ten or twenty thousand people in the head office and R&D alone. No matter how you look at it, this is way too many.

Underlying this situation is that, in the midst of the postwar emphasis on academic achievement, white-collar workers rose in number and administrative staffs ballooned. One benefit was that toward the end of the 1980s many highly educated people were assigned to manufacturing sites and sales positions, increasing the level of staff. This emphasis on education created a high level of company culture in Japan and produced results that would astound the world.

But even after the period of high economic growth ended, back-office staff continued to increase at the same overblown pace, until today its size has grown out of all proportion to the value added. Administrative staff has continued to grow, and since many of these people are highly paid, the resulting personnel costs have become a burden on Japanese companies. This in turn has put pressure on overall profitability and is one reason that the current productivity of the Japanese workforce is so low.

According to media sources, Japanese productivity, measured as GDP divided by the total workforce, is only about two-thirds that of the United States. This is a serious problem that Japanese companies will have to tackle in the near future.

Issues for Japan: Deterioration in the Ability to Execute

One more thing that concerns me about Japanese business is the deterioration in the ability to execute. This can be referred to as a front-line weakness. The problem we face is how to upgrade our front-line forces.

Japanese soldiers up to the noncommissioned officer level are superlative in quality. Added to this is the deeply ingrained ability of the Japanese to persevere under difficult circumstances. However flawed the strategy of a commissioned officer, the front-line foot soldiers have been always able to compensate by their sheer ability to get the job done.

I can't help thinking that recently this Japanese ability to execute is not quite what it used to be. My overall impression is that the energy thrown into getting a project done is sorely lacking. This criticism applies not only to Fujifilm but to all Japanese companies; in fact, to all Japanese.

Let's say that the target set for a particular period is not met. It is one thing if this target has been determined by strong-arm managers who are ignorant of on-site conditions, but in most cases the target figures have been calculated on the basis of numbers supplied by the various business units themselves. If the targets coming from on-site managers cannot be met, you have to conclude that something is wrong with on-site execution. This deterioration in the ability to get the job done is much talked about among our corporations, and is definitely one of the problems with Japanese industry that must be rectified in the future.

How can the level of execution be raised at each workplace and by each individual? Companywide education might be effective to a certain extent, but each location has to set practical goals and strive to achieve them one at a time with all the energy and intellect they can muster.

Fujifilm inaugurated a new regime in 2012, and one of its goals is simply to improve the ability to execute.

Issues for Japan: Blurring Responsibility

One more worrisome issue concerns Japan as a whole. When something has gone wrong, there is a tendency to forgo proper analysis and investigation, to refuse to face reality, and to blur the lines of responsibility. This can be said both of corporate management and of political leaders alike.

The most egregious example of this is Japan in the postwar years. After being defeated in World War II, the country should have carefully scrutinized and sincerely reflected on the reasons why it went to war and why it lost. Instead, Japan rationalized the conflict by saying the army had gotten out of control, and refused to engage in any serious reflection. Such a self-examination should have received top priority. Questions should have been asked: What exactly went wrong? What changes does Japan have to make? How do the Japanese themselves have to change? What should be preserved and what abandoned? Had such questions been asked and answered, lessons for the future would have been learned.

The same can be said about the Fukushima Daiichi Nuclear Power Station incident that followed in the wake of the Great East Japan Earthquake and tsunami in March 2011. How could such a catastrophe occur? Although the matter should have received the closest study, it still awaits that scrutiny today.

Strategic errors and policy miscalculations are bound to happen in business management, too, and in Japan the tendency is to look the other way. Slowly but surely the problem is enveloped in a cloud of smoke, which essentially means that no one takes responsibility.

I can understand the feelings of sympathy that Japanese have for the defeated and for those who have made mistakes. To have such feelings may be a virtue. But if you've screwed up or been on the receiving end of one, you are likely to repeat the experience if you don't clarify what went wrong and establish responsibility.

On that count Japanese society is still morally ambiguous. To compete successfully on the global stage, Japan as a whole must learn to recognize a fact as a fact, to face reality, and to do so with courage.

Japanese Technology: Still a Source of Pride

While problems remain, I am firm in my belief that Japanese companies have great potential. The ultimate strength of corporate Japan is technology. This technology is sustained by hardworking, loyal employees, who are always striving to make something better, something new. Even at Fujifilm, the constant thought is to move forward, to always keep moving forward. In a great many industries, Japanese companies are constantly looking toward the future, always thinking in terms of the cutting edge. They are not satisfied with the easy and the cheap.

With its rapidly advancing technology, Japan still has a big lead in most of the high-tech fields. I am often frustrated by critics who say that Japan's manufacturers are falling behind. These critics don't understand the real strength of Japanese companies, their driving energy, and their attitude toward work. We shouldn't fear the future unless we have reason to fear the future. If we continue to improve on the advanced technology we now have, and wed that technology to superlative products, there is no reason to be overly concerned about what will happen to Japan.

Japan's strength is evident not only in single technologies, but also in cases where several leading-edge technologies are combined. The liquid crystal display is a Japanese innovation, but it was not created by one company alone. It was the product of a number of companies—optical film makers like Fujifilm, glass makers, semiconductor manufacturers, and electronics assembly companies—each contributing its technical expertise to create a product that was feasible in terms of quality and price, a product that could serve as another example of Japan's industrial prowess.

In fact, our industrial prowess—the technological strength of Japan's companies—hasn't undergone any great change, and it should remain a strong economic weapon into the future. A much finer global perspective is necessary, however: What new products should be created? How should we fit into the global market? How do we compete with industries from different fields?

Strategic products, greater efficiency through a variety of systems and innovations, and accelerated business processes are more crucial than ever, and form a large part of the answer. For all these to come about, upgrading the quality of personnel will be indispensable, along with maintaining and lifting up the basic education level of the Japanese people. This is precisely why the Japanese educational system will become an important topic for discussion in the coming days.

Teaching Children the Importance of Competition

You cannot raise the standard of Japanese technological and industrial strength or upgrade the level of strategic planning unless the people entrusted with the task are properly trained and educated. Greater emphasis on teaching young students the importance of competition is especially important.

Competition is an essential part of life. Living in Japan one may not be strongly aware of it, but the rest of the world revolves around this principle. With the advent of a borderless world, where competing on a global level is a matter of course, everyone will have to become accustomed to competition at an early age in order to survive.

Look at the education Japan is now providing its children. Quarreling is not allowed. Keeping track of who finished in which place in foot races is not done. Friction of any kind—conflict, discord, strife, what have you—is avoided. Anything that smacks of competition is ruled out. As a result, we are producing a lot of gentle, pliable, polite children, which is fine, I suppose. But can these children compete in a fiercely dog-eat-dog world? Do they even have the desire to compete?

> *Competition is harsh, but it makes people grow. We can't afford to wait: there is an urgent need to create the conditions for fostering people who are mentally tough.*
>
> *Competition also makes it possible to know yourself. Recently the empty expression "I am looking for myself" has reached my ears, but I don't know how in the world can people find themselves in the absence of struggle and competition.*
>
> *Competition isn't just about being a good student or good at sports. Even if you are deficient at studies and athletics, if you*

can make people laugh, you can compete at that and perhaps become a star entertainer. Some people like singing, some like drawing; everyone has something they are good at. It's impossible for everyone to be the same.

Competition allows you to discover where you excel, and to find out what you like. Because of competition you can find the place where you belong. You begin to see the differences in individuals, and how each individual is able to live a life most suited to him- or herself. This is how society maintains its balance.

By removing competition and trying to make all people equal, everyone ends up going in the same direction. What is left is a life without variation, where you are no different from anyone else and can't express your individuality—a boring, lackluster life.

Those who have managed to reach a coveted position in society are not necessarily the winners in life. Those who try their best, who direct their own lives, who express their individuality—they are the true winners.

From the beginning, every person is different. By recognizing how you are different from others, by knowing yourself, you can come to terms with life.

When you try to deny this truth, you end up getting discouraged at the slightest thing, and come to the mistaken conclusion that, after all, it might be better to live like everyone else.

Education, thus, should be geared to the individual. Avoiding competition and trying to make everyone possess the same abilities and produce the same results is wrong from the start.

I am fully in favor of elite education, even of extending and enhancing it. Skipping grades is fine. Targeted education for leaders is fine. This is how it is normally done abroad, and if as a Japanese you go overseas you will instantly find yourself duking it out with these elite businesspersons with their Harvard and Wharton degrees.

Overseas, the easygoing idea that "We're all buddies; we don't compete among ourselves" will get you nowhere. Try giving a fake laugh and you'll only be looked down on. In our global age, true communication is established by looking your opponent squarely in the eye and engaging in hard-nosed competition. Make no mistake about it; the business world is a battlefield.

In populous and rapidly growing countries like China and India, competition is incredibly fierce. Children even flunk out of elementary school in some of these countries,. That's why they study so hard and try not to fall behind. As a result, academic standards there have skyrocketed. The International Mathematical Olympiad used to be Japan's one-man show, but today Japan can hardly compete. If Japan continues its noncompetitive, anodyne approach to education, I fear for the future of our country.

From Backward, Inward, and Downward to Forward, Outward, and Upward

There was a time in the past when Japan struggled mightily to achieve equal status with the Western powers. From the Meiji period in the late nineteenth and early twentieth centuries to the subsequent Taisho and early Showa periods, the national strength of the country expanded at a dramatic pace. What lay behind this urgency was the fear that Japan would be colonized by the West in much the same way that nearby countries were being colonized.

So the Japanese people set to work body and soul. With the goal of catching up to the West, they gritted their teeth and worked their fingers to the bone. The effort they expended deserves the highest respect. They were driven by the fear that the nation itself was in danger, and it was this fear that drove the growth of the country.

Do the Japanese people today have the spirit of the Japanese of that period? I hear voices lamenting that the country has been left behind. Young people say that Japan is finished, but I believe such talk is absurd. The spirit of refusing to lose, the sense that we'll come out on top, should never be forgotten. But if things continue as they are today, I sincerely feel that Japan's position is precarious.

But I am not entirely pessimistic about Japan's future. If the strengths of Japan and the Japanese people are brought fully into play, there is ample chance for victory. Confronting the up-and-coming Asian countries, challenging the aggressive West and its trade tactics, Japan needs now to return to the enterprising spirit of the Meiji period, to that spirit and energy, to build a twenty-first century Japan.

I've said a great deal here about the Japanese economy, but in the end the solution is actually quite simple. It is just a matter of having the right spirit, of refusing to lose, refusing to give up. From personal experience I have seen the Japanese people rise up again and again after being plunged into the depths of despair.

In 1945, after losing the war, Japan had essentially been beaten to the ground. Almost all of our major cities had completely ceased functioning. Most Japanese had lost everything: their homes, their companies, and their work. But Japan rose to its feet.

Progress is never in a straight line. Yet despite all the ups and downs, Japan remains a technological front-runner. Its people possess a laudable work ethic, loyalty to a worthy cause, a sense of responsibility, and a strong feeling of solidarity. If these mental pillars of strength are reconfirmed, the country's forces remarshaled, and the battle joined in earnest, there is no reason Japan should not be able to compete at the highest level. With the spirit of "Refusing to Lose," Japan must once again become a global challenger.

I often speak of the importance of the "Three Orientations": forward, outward, and upward. Japan today has them turned around to: backward, inward, and downward, having thoroughly lost confidence in itself. Nevertheless, Japan is still a supremely powerful industrial country. It has a trove of assets in technology, people, and money, and it still has the wherewithal to win. Throughout the world, every nation has its own particular problems. The United States is no exception, and Japan is better off than most. What Japan now needs to do is hit the restart button and reset its collective mind.

This is not the time to be saying Japan is finished. The younger generation has to recognize its role in creating a new Japan. The future of the country is in their hands. The Japanese nation—the Japanese people—need to rediscover their pride, to revitalize their fighting spirit, and to redirect their thinking to a forward, outward, and upward orientation. Everything—everything—starts with the recovery of the Japanese fighting spirit.

CONCLUSION

In the city of Suzhou, China, the Fujifilm group has four factories. Nearby is the Hanshan (Cold Mountain) Temple, which is well known for its mention in a Tang-dynasty poem by Zhang Ji. The poem is called "A Night Mooring by Maple Bridge" and reads as follows:

> *The old moon goes down; the crows make a ruckus. The world is covered with frost.*
>
> *There are maples on the riverbank, and fishing-boat lights drift on the current.*
>
> *I fall into a sad sleep, and from the monastery on Cold Mountain*
>
> *The sound of a bell reaches the guest boat at midnight.*

The head priest, Qiu Shuang, was a very quiet man, and when he once offered to write something for me, I immediately assented. The calligraphy he inscribed at my request then is hanging today in the reception room of our head office.

The word I requested was "courage," which in

Japanese consists of two characters. I think of courage as the bedrock of my value system and my life itself. When you are required to make a decision, you must believe in your own judgment and make it with courage.

The first of the two characters means "bravery," and its shape reminds me of a square-shouldered man overflowing with courage and determination. The second means "spirit," and it seems to fill the air with heroic valor. This calligraphy is one of my favorites.

In the Advanced Research Laboratories built in April 2006 as part of the reforms at Fujifilm, there is a special symbolic image—a representation of the Roman goddess Minerva with an owl. In the preface to his *Elements of the Philosophy of Right*, G. W. F. Hegel famously writes, "The owl of Minerva begins its flight only with the onset of dusk."

Minerva, in Roman mythology, was the goddess of crafts and war and the personification of wisdom. The owl was her sacred bird. When a civilization or an epoch came to an end, Minerva would release the owl and have the large-eyed bird appraise the age—what kind of era was it, how did it come to an end? Then, with this appraisal in hand, she would prepare for the age to come.

In Fujifilm's case, the company's near twilight came with the end of its peak years producing photosensitive materials, and the forward-looking research center was to serve as a base camp for the creation of new core technologies. It was to serve the purpose of

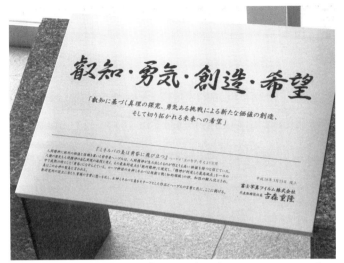

Plaque at Fujifilm Advanced Research Laboratories, engraved with the words "Wisdom, Courage, Creation, Hope."

Minerva's owl. On this monument I had these four words engraved: Wisdom, Courage, Creation, Hope. The message was to search for the truth based on wisdom; to have the courage to create new values; to have high hopes for the imminent future. The word "courage" could not be ignored.

The Fujifilm reforms were not, of course, instigated by me alone. The company already had a firm foundation and a sturdy backbone, thanks to everyone's hard work in establishing an enterprise boasting high-level technology and solid finances. I thank both senior and junior colleagues for their work in building a corporate foundation capable of withstanding adversity. The reformation of Fujifilm was only possible because each individual, backed by the company's fundamental strength, responded wholeheartedly in his or her own way. I hope it was my leadership that inspired them. To all these people I express my thanks and appreciation. It is my fervent hope that with their unswerving efforts, Fujifilm will continue to grow into the future.

Fujifilm is now attempting to go beyond its established business of imaging and information and transform itself into a company that can make a wider contribution to science, culture, healthcare, and the preservation of the environment. Fujifilm's philosophy is as follows:

> *We will use leading-edge, proprietary technologies to provide top-quality products and services that contribute to the advancement of culture, science, technology, and industry, as well as improved health and environmental protection in society. Our overarching aim is to help enhance the quality of life of people worldwide.*

To maintain its position as a top-flight company and play an indispensable social role throughout the twenty-first century, Fujifilm must be prepared to soar to new heights. Using its proprietary technology and rich experience, it must continue to provide innovative values to the world. I would like to see Fujifilm move forward through the concentrated knowledge and determination of each and every employee, and thus fulfill the company's noble and inspirational goals.

* * *

This is my first book. I have tried to give a faithful account of how I reacted as the top leader when Fujifilm was faced with a dramatic change in its environment: what I thought, what decisions I made, how I implemented those decisions.

The changes now taking place in the corporate environment are accelerating at an exponential rate. These dramatic changes are not confined to Fujifilm, but are a challenge shared by a great many enterprises. If through this book I have been able to convey something of the courage with which Fujifilm confronted these problems, I will be immensely pleased.

Innovating Out of Crisis: How Fujifilm Survived (and Thrived) As Its Core Business Was Vanishing

© 2015 SHIGETAKA KOMORI

Published by STONE BRIDGE PRESS, P.O. Box 8208
Berkeley, CA 94707

No part of this book may be reproduced in any form
without written permission from the publisher.

発売元　IBCパブリッシング株式会社
〒162-0804　東京都新宿区中里町 29 番 3 号
菱秀神楽坂ビル 9F　Tel. 03-3513-4511

ISBN 978-4-7946-0340-1